# Collection of Solved Problems in Vibration

Dragi Tiro

Protea Publishing

Collection of Solved Problems in Vibration
by Dragi Tiro

First Edition Wordlwide

ISBN 1-931768-42-0

Protea Publishing
Norcross, Georgia, USA

email: kaolink@msn.com

web site: www.proteapublishing.com

Library of Congress Control Number: 2002103266

## Acknowledgements

I wish to acknowledge the contributions of many people. Thanks to reviewers of the manuscript:
Stefan Sunaric, Ph.D.
and Alija Picuga, Ph.D.
I appreciate the excellent cooperation and facilities of Protea Publishing Company and Johan du Toit.
I wish to take this opportunity of expressing my appreciation to Mr. Cory Anibal for the book proofreading.
Finally, a note of gratitude also goes to my family for their patience, especially to my father Arif Tiro for helpful comments about the printing business.

Dragi Tiro

## Review of Contents

# Contents

# CHAPTER ONE

## INTRODUCTION

In this book we describe the process of solving problems of small linear vibrations of mechanical systems. The solving process for all problems has three phases as shown in Figure 1.1.

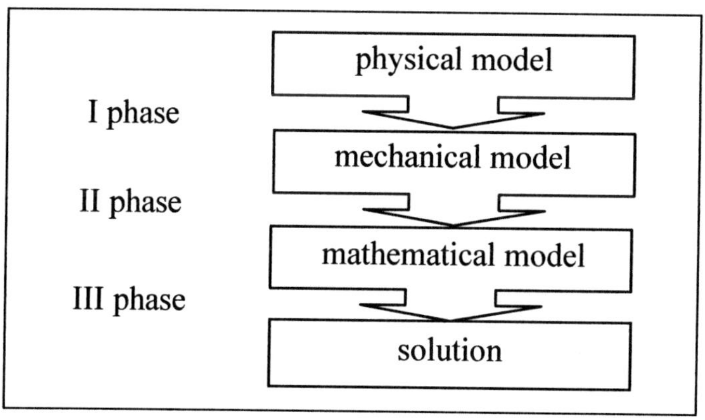


*Figure 1.1. The problem solving process has three phases*

First phase: Draw a sketch of the mechanical model. While the real mechanical system is physical model with a real system of vibrations, we must make some approximations to render a simpler mathematical model. For instance: In one physical model we would focus on those parts where mass is concentrated and disregard areas of minor mass values. In another instance we could consider some masses as absolute rigid and disregard areas with minor resistance to motion. In this way it is possible to consider elastic part as a spring. One might even reduce closely equivalent masses into one. These approximations do not influence the final solution significantly.
Second phase: Apply the laws of dynamics to the mechanical model and find the differential equations for

vibrations. The physical model now becomes a mathematical model.
Third phase: Calculating the physical properties of vibrations (such as circular frequency $\omega$ or period $T$) from the mathematical model and the law of vibrations gives the position of the vibration system as a function of time.
Physical units are stated in square brackets in this book.

*Problem: Figure 1.2 shows the estimation of a pendulum's position by the law of mathematical pendulum vibrations,* $\varphi = 0{,}2 \cdot \cos(31{,}3 \cdot t)$*. Estimate the position of the pendulum at moments:* $t_1=0[s]$, $t_2=1[s]$, $t_3=3[s]$.

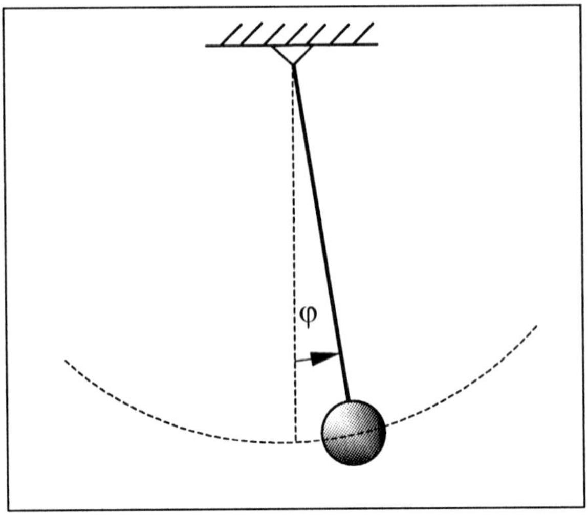


*Figure 1.2. The mathematical pendulum*

*Solution:*
*The position is decided with angle $\varphi$ at moment $t_1$:*

$$\varphi_1 = 0{,}2 \cdot \cos(31{,}3 \cdot 0) = 0{,}2[rad]$$

*Moment $t_2$:*

$$\varphi_2 = 0{,}2 \cdot \cos(31{,}3 \cdot 1) = 0{,}17[rad]$$

*Moment $t_2$:*

$$\varphi_3 = 0{,}2 \cdot \cos(31{,}3 \cdot 3) = -0{,}014[rad]$$

Mathematical models in this book are solved in the classical way without the use of a computer. There are many software tools for solving different mathematical forms, off-the-shelf products like Mathematica®, MATLAB®, MAPLE® and others.
A short review of established theory opens each of the following nine chapters. Knowledge of this theory is necessary for the reader to follow the solution of the problems that follow. And, the solutions make the theory quite clear. The problems in each chapter are arranged in increasing degree of difficulty.
Focus is on small vibrations, i.e. the small displacement of the vibration system round the position of equilibrium. The reader should have a basic knowledge of: statics, kinematics, dynamics, mathematics, and understand, for example, conditions of equilibrium, angular speed, the laws of dynamics, inertial force, and differential equations.
The problems are solved using laws of dynamics such as *the Law of Change of Angular Momentum*, *d'Alembert's Law, the Newton's Second Law (Basic Law) of Dynamic.*
**The mass of the spring is disregarded in all the problems.** The spring is shown in Figure 1.3.

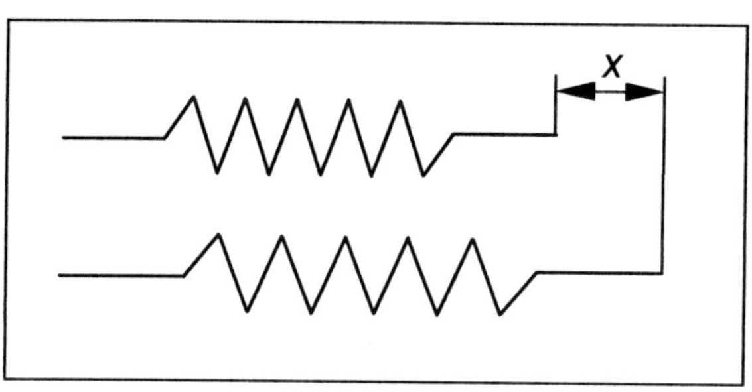


*Figure 1.3. The spring (x is extended length of the spring)*

Force of the spring:
$F = c \cdot x$ ,
where $c$ is the rigidity of the spring and x is the extended length of the spring (or the compression of the spring). It is possible to observe a bar like the spring (Figure 1.4.).
The Hooke's law:

$\sigma = E \cdot \varepsilon$ ,

where $\sigma$ is the intensity of stress, $E$ is Young's modulus of elasticity and $\varepsilon$ is the dilatation. The intensity of stress: $\sigma = F/A$, where $F$ is the force in the bar and $A$ is the surface of bar's cross section. Now, we can write the Hooke's law as:

$$\frac{F}{A} = E \cdot \frac{\Delta l}{l}$$

$$F = \frac{E \cdot A}{l} \Delta l$$

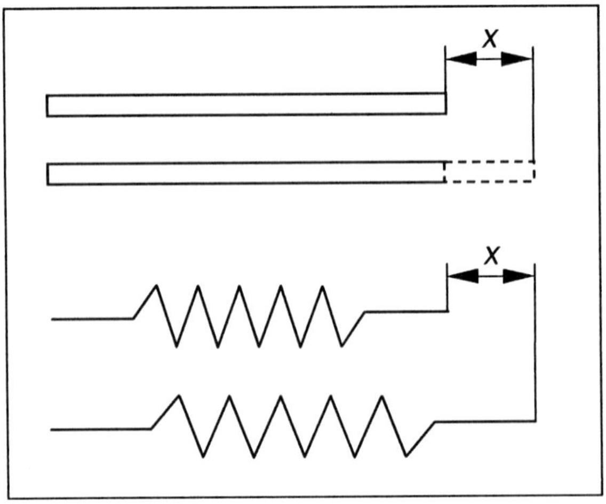


*Figure 1.4. It is possible to observe a bar like the spring*

The rigidity of the bar:

$$c = \frac{E \cdot A}{l}$$

$F = c \cdot \Delta l$ or $F = c \cdot x$

So, the force in the bar is a product of the rigidity and the extended length of the bar (or compression of the bar). This is valid when the intensity of stress is less than the limit of elasticity.

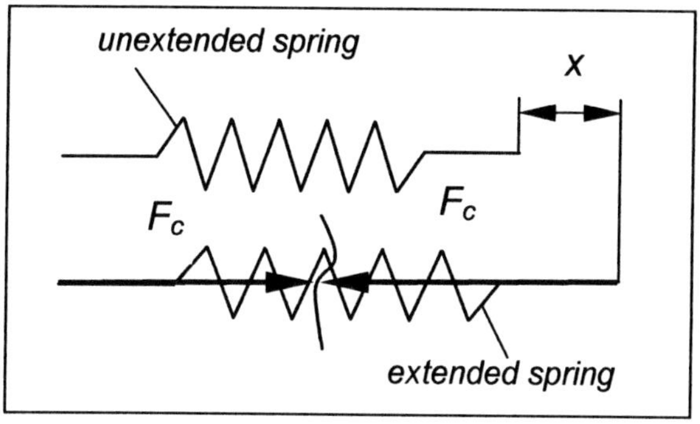


*Figure 1.5. The force of extended spring*

Imagine that we cut the spring. If the spring is extended, the force of spring has sense towards the section (Figure 1.5.). If the spring is compressed, the force of spring has sense from the section (Figure 1.6.).

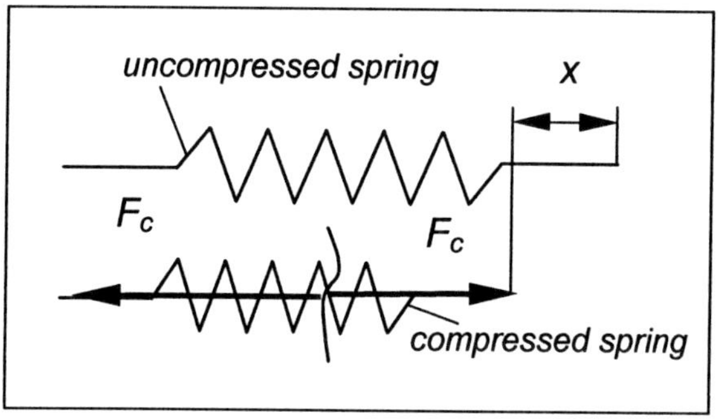


*Figure 1.6. The force of compressed spring*

# CHAPTER TWO

## FREE UNDAMPED VIBRATIONS WITH ONE DEGREE OF VIBRATION'S FREEDOM

### 2.1. Short Theory Review

It is necessary to draw a sketch of vibration's system displaced off from the balance position, and then to find a differential equation with using the some of dynamical laws. The found differential equation characterises the vibration's system.

This kind of vibrations is described with differential equation form:

$$\frac{d^2q}{dt^2} + \omega^2 q = 0 \qquad \text{.....................(2.1)}$$

It's a general form of the differential equation for free undamped vibrations with one degree of vibration's freedom.

In equation (2.1.) the marks are:

*q* - generalized coordinate (generalized means that the coordinate can be linear or angular),

*t* - time,

$\omega$ - circular frequency.

Solution of differential equation (2.1.) has the form:

$$q = e^{rt} \qquad \text{.........................(2.2.)}$$

If we put the equation (2.2.) in (2.1.) and set it in order [1], we will get the law of vibrations:

$$q = C_1 \cos\omega t + C_2 \sin\omega t \qquad \text{..........(2.3)}$$

where $C_1$ and $C_2$ are the constants and it is possible to calculate them from initial conditions (starting position and initial velocity).

Solution (2.3.) can get another one form. If we put:

$$C_1 = A \sin\alpha \qquad \text{.........................(2.4.)}$$

$$C_2 = A \cos\alpha \qquad \text{.........................(2.5.)}$$

in (2.3.), we will get:

$$q = A \sin(\omega t + \alpha), \qquad \text{...................(2.6.)}$$

where $A$ is amplitude, $\alpha$ is initial phase, $(\omega t + \alpha)$ is a phase of vibrations. $A$ and $\alpha$ are constants, and they depend upon initial conditions.
Period or time of swings $T$:

$$T = {}^{2\pi}\!/_{\omega} \quad [s] \qquad \ldots\ldots\ldots\ldots\ldots\ldots (2.7.)$$

is the time of one vibration.
$T$ and $\omega$ depend upon vibrations system characteristic (mass, rigidity, ...). Frequency of vibrations $\nu$:

$$\nu = \frac{1}{T} = \frac{\omega}{2\pi} \quad [Hz] \qquad \ldots\ldots\ldots\ldots\ldots\ldots (2.8.)$$

is the number of vibrations for one second.
Relatively the most simple and quick way for getting the differential equation is using of LaGrange's equation. For free undamped vibrations with one degree of vibration's freedom the equation takes this form:

$$\frac{d}{dt}\left(\frac{\partial E_k}{\partial \dot{q}}\right) - \frac{\partial E_k}{\partial q} + \frac{\partial E_p}{\partial q} = 0 \qquad \ldots\ldots\ldots (2.9.)$$

where $E_k$ is kinetic energy of vibration's system, $E_p$ is potential energy of vibration's system, $q$ is generalized coordinate and $\dot{q} = \frac{dq}{dt}$ is the first derivative of coordinate $q$ with respect to time $t$. $\frac{\partial \cdots}{\partial \cdots}$ is the mark for partial derivative (for example $\frac{\partial E_p}{\partial q}$ is partial derivative of potential energy with respect to coordinate $q$).

## 2.2.Problems

2.1. Concentrated mass $m$ vibrates inside a pipe (see Figure 1.1.) and the mass is linked for an immovable wall with a spring. The spring has rigidity $c$. Find differential equation of vibrations, circular frequency $\omega$, period $T$, frequency of vibrations $\nu$ and movement law. It is known: $m$=2,5[kg];

$c$=16,92[N/cm]; $\alpha$=30°. Initial conditions: $t_0$=0 [s], $x_0$=0 [m], $\dot{x}_0 = 0{,}5\left[\frac{m}{s}\right]$.

Friction between the mass and the pipe is neglected.

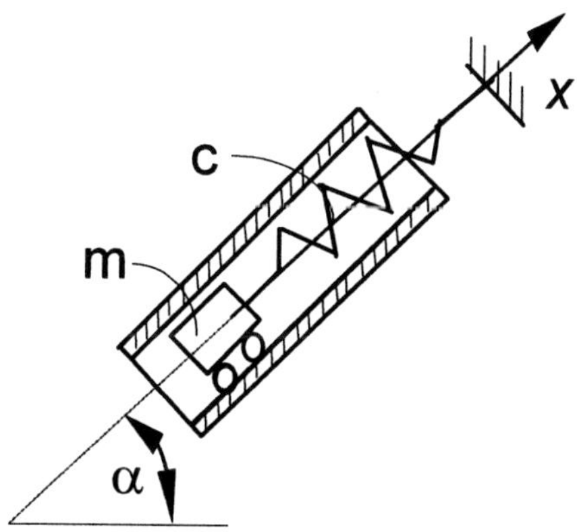


*Figure 2.1.*

Solution:

Vibrations of the system are linear (the mass moves just in axis $x$ in Figure 2.1.), so the Newton's Second law is the most favourable for solving the problem. It is possible to determine a position of mass with coordinate $x$. It means there are the vibrations with one degree of vibration's freedom.

Length of the spring without load is $l$. The spring will be stretched for $f_{st}$, when we put on the mass. Static force $F_{st}=c\cdot f_{st}$ is appeared in the spring. Static condition of equilibrium:

$$F_{st} = G\cdot \sin\alpha$$

$$cf_{st} = mg\sin\alpha \qquad (2.2.1.)$$

If we displace the mass off the position of equilibrium and let it go, the mass will vibrate about the position, by reason of the spring's force action. Newton's Second law:

$$m\ddot{x} = \sum F_x$$

$$m\ddot{x} = G\sin\alpha - F_c \qquad (2.2.2.)$$

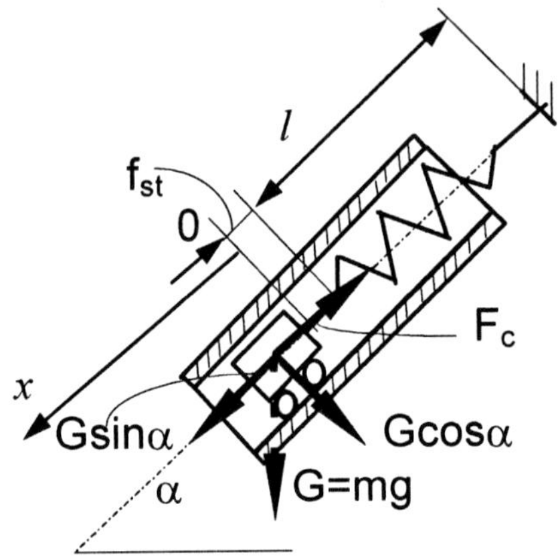


*Figure 2.2.*

*Fc* is the force in spring, and *Fc=c· (fst + x)*. If we put it in the equation (2.2.2.), we will get:

$$m\ddot{x} = G\sin\alpha - c\cdot\left(f_{st} + x\right)$$

$$m\ddot{x} = G\sin\alpha - c\cdot f_{st} - c\cdot x$$

$$m\ddot{x} + c\cdot x = m\cdot g\cdot\sin\alpha - c\cdot f_{st} \quad \text{(2.2.3.)}$$

Put the equation (2.2.1.) in (2.2.3.):

$$m\ddot{x} + c\cdot x = 0$$

$$\ddot{x} + \frac{c}{m}\cdot x = 0 \quad \text{(2.2.4.)}$$

The equation (2.2.4.) is differential equation of free undamped vibrations with one degree of vibration's freedom. With comparing the equations (2.1.) and (2.2.4.) we get:

$$\omega^2 = \frac{c}{m}$$

$$\omega = \sqrt{\frac{c}{m}} = \sqrt{\frac{16{,}92\cdot 100\frac{N}{m}}{2{,}5kg}} = 26\left[\frac{rad}{s}\right]$$

The period:

$$T = \frac{2\pi}{\omega} = 2\pi\sqrt{\frac{m}{c}} = 2\cdot 3{,}14\cdot\sqrt{\frac{2{,}5kg}{16{,}92\cdot 100\frac{N}{m}}} = 0{,}24[s]$$

Frequency of vibrations $\nu$:

$$\nu = \frac{1}{T} = \frac{\omega}{2\pi} = \frac{26\frac{rad}{s}}{2\pi} = 4{,}138\ [Hz]$$

It is possible to suppose the solution of the differential equation (2.2.4.) is in the form (2.3.):

$x = C_1 \cos \omega t + C_2 \sin \omega t$

$$\frac{dx}{dt} = \dot{x} = -C_1 \omega \sin \omega t + C_2 \omega \cos \omega t$$

Put the initial conditions:

$x_0 = C_1 \cos \omega t_0 + C_2 \sin \omega t_0$

$$0 = C_1 \cdot 1 + C_2 \cdot 0 \quad \Rightarrow \quad C_1 = 0[m]$$

$$0{,}5 = -C_1 \cdot 26 \cdot 0 + C_2 \cdot 26 \cdot 1 \ \Rightarrow 0{,}5 = 26 \cdot C_2 \ \Rightarrow C_2 = \frac{1}{52}[m]$$

The law of vibrating:

$$x = \frac{1}{52} \sin 26 \cdot t \qquad [m, s]$$

2.2. Estimate the circular frequency $\omega$ and the time of swing $T$ for vibration's system in Figure 2.3. It is known: $m$=2,7 [kg], $c$=2000 [N/m], $\alpha$=45°. Friction between the mass and the pipe is neglected.

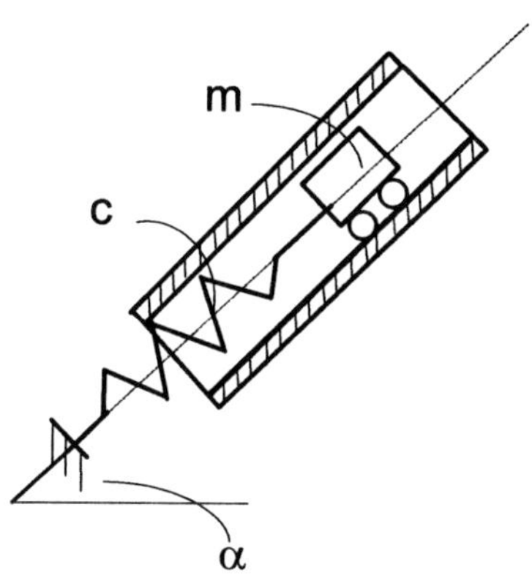


*Figure 2.3.*

Solution:

Static condition of equilibrium (Figure 2.4.):

$$\sum F_{ix} = 0$$

$G_x$-$F_c$= 0,

G $\sin\alpha$ - $cf_{st}$= 0 ,

$mg \cdot \sin\alpha$ -c $f_{st}$= 0 . . . . . . . . . . . . . . . . . (2.2.5.)

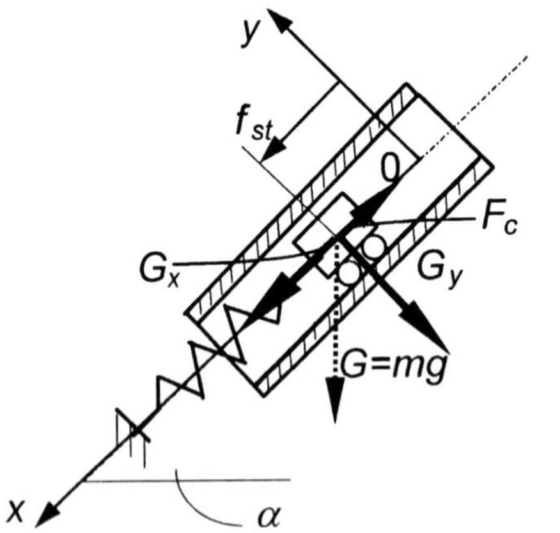


*Figure 2.4.*

The mass is displaced from the position of equilibrium in Figure 2.5. Newton's Second law for the system in the Figure:

$$m\ddot{x} = \sum F_{ix} \quad \Leftrightarrow \quad m\ddot{x} = G\sin\alpha - F_c$$

$$m\ddot{x} = G\sin\alpha - c\cdot(f_{st} + x)$$

If we put equation (2.2.5.) in the previous one, we will get:

$$m\ddot{x} + cx = 0 \,/_{:m}$$

$$\ddot{x} + \frac{c}{m}\cdot x = 0 \quad \Rightarrow \quad \omega^2 = \frac{c}{m} \quad \Rightarrow$$

$$\omega = \sqrt{\frac{c}{m}} = \sqrt{\frac{2000\frac{N}{m}}{2{,}7\ kg}} = 27{,}217\left[\frac{rad}{s}\right]$$

$$T = \frac{2\pi}{\omega} = \frac{2\cdot 3{,}14}{27{,}217\frac{rad}{s}} = 0{,}23[s]$$

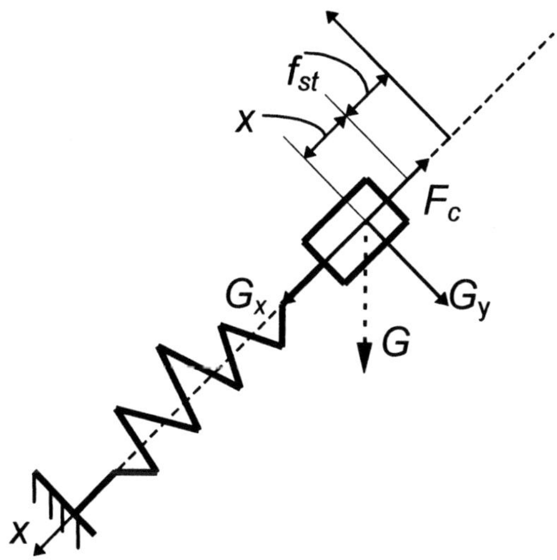


*Figure 2.5.*

2.3. Concentrated mass *m* is placed at the end of a lever. Mass of the lever is neglected. A spring is hitched on the lever at the distance *a* from the articulated coupling *O* (Figure 2.6.). The rigidity of the spring is *c*. The lever rotates around point *O* without friction. Estimate a differential equation of movement by using d'Alembert's law, then determine physical characteristics (period *T* and circular frequency *ω*) and the law of movement.

The initial conditions are: $t_0=0$; $\varphi_0=0$; $\left(\frac{d\varphi}{dt}\right)_0 = 0{,}3\left[\frac{m}{s}\right]$ .

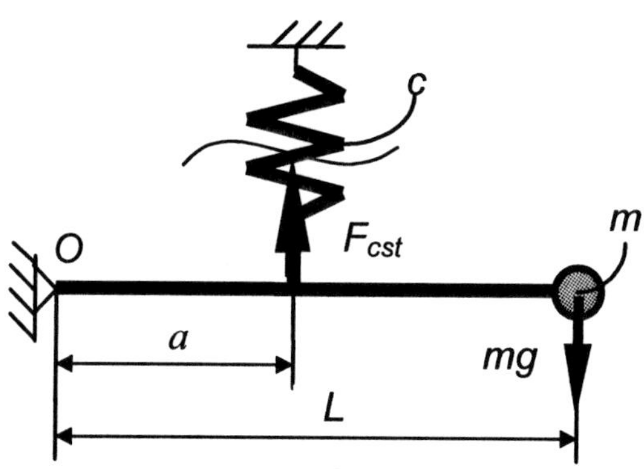


*Figure 2.6.*

Solution:
Static condition of equilibrium (Figure 2.6.):

$$\sum M_0 = 0$$

$$F_{cst} a - GL = 0; \quad cf_{st} a - mgL = 0$$

It is necessary to displace the system off the position of equilibrium and to add the inertial forces for every mass. The system has one mass (the others are neglected) and one inertial force, as shown in Figure 2.7.

The inertial force: $F_T^{in} = mL\ddot{\varphi}$

Normal inertial force isn't necessary for us, because its direction passes by point O, and hasn't torque for the point O:

$$\sum M_O = 0$$

$$F_T^{in} \cdot L - mgL\cos\varphi + F_c a\cos\varphi = 0$$

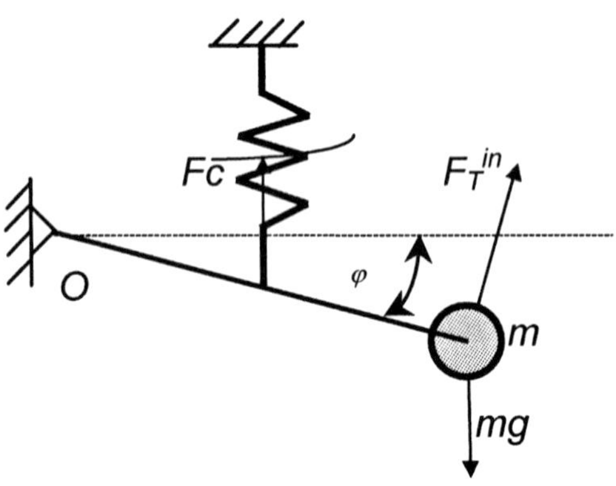


*Figure 2.7.*

For small vibrations we always approximate sin$\varphi$ with $\varphi$, and cos$\varphi$ with 1. With this approximation we get a simpler equation. Also, we use the static condition of equilibrium:

$$mL^2\ddot{\varphi} - mgL \cdot 1 + c \cdot (fst + a \cdot \varphi) \cdot a \cdot 1 = 0$$

$$\ddot{\varphi} + \frac{a^2 c}{mL^2}\varphi = 0 \Rightarrow \omega = \frac{a}{L}\sqrt{\frac{c}{m}}; T = \frac{2\pi}{\omega}$$

$$\omega = 15\left[\frac{rad}{s}\right] \quad ; \quad T = 0{,}42[s]$$

The law of vibrations:

$$\varphi = A\cos\omega\, t + B\sin\omega\, t$$

$$\frac{d\varphi}{dt} = -\omega\, A\sin\omega\, t + B\omega\cos\omega t$$

The initial conditions: $t_0=0$ ; $\varphi_0=0$ ; $\left(\frac{d\varphi}{dt}\right)_0 = 0{,}3\left[\frac{m}{s}\right]$ are inserted in two equations above:

$$0{,}3\frac{m}{s} = -\omega A\sin\omega\cdot 0 + \omega B\cos\omega\cdot 0 \quad \Rightarrow$$

$$\Rightarrow \qquad B = \frac{0{,}3\frac{m}{s}}{\omega} = \frac{0{,}3\frac{m}{s}}{15\frac{rad}{s}} = 0{,}02[m]$$

$$0 = A\cos\omega\cdot 0 + B\sin\omega\cdot 0 \quad \Rightarrow \quad A=0[\mathrm{m}]$$

The law of vibrations:

$$\varphi = 0{,}02\cdot\sin 15\cdot t \quad [m,s]$$

2.4. Solve the problem 2.3. using LaGrange's equation.

Solution:

LaGrange's equation for free undamped vibrations with one degree of vibration's freedom has form (2.9.):

$$\frac{d}{dt}\left(\frac{\partial E_k}{\partial\dot{q}}\right) - \frac{\partial E_k}{\partial q} + \frac{\partial E_p}{\partial q} = 0$$

It is necessary to estimate kinetic and potential energy of the vibration's system. Potential energy of the spring is marked with $E_{pc}$, and of the mass with $E_{pm}$. Potential energy of the spring is always:

$$E_p = \frac{1}{2}c\cdot x\,,$$

where c is rigidity of the spring, and $x$ is extended length of the spring (Figure 1.3.). For vibration's system in figure 2.7:

$$E_p = E_{pc} + E_{pm}; E_{pc} = \frac{1}{2}c(f_{st} + a\cdot\sin\varphi)^2$$

For small vibrations: $\sin\varphi \approx \varphi$

Potential energy has negative value if the mass goes down. The mass goes down in Figure 2.7.

$$E_{pm} = -mgh; h = L\cdot\sin\varphi \approx L\varphi$$

$$E_{pm} = -mgL\varphi; E_p = \frac{1}{2}c(f_{st} + a\varphi)^2 - mgL\varphi$$

$$E_k = \frac{1}{2}mv_m^{\ 2} = \frac{1}{2}L^2\dot{\varphi}^2$$

$$\frac{\partial E_k}{\partial\dot{\varphi}} = 2\cdot\frac{1}{2}mL^2\dot{\varphi}; \quad \frac{\partial E_p}{\partial\varphi} = 2\cdot\frac{1}{2}c\cdot(f_{st} + a\varphi)\cdot a - mgL; \quad \frac{\partial E_k}{\partial\varphi} = 0$$

$$mL^2\ddot{\varphi} - 0 + c\cdot a\cdot(f_{st} + a\varphi) - mgL = 0$$

$$mL^2\ddot{\varphi} + c\cdot a\cdot f_{st} + c\cdot a\varphi - mgL = 0$$

The static condition of equilibrium (Figure 2.6.):

$$\sum M_0 = 0$$

$$F_{cst}a - GL = 0; \quad cf_{st}a - mgL = 0$$

It is needed to insert the condition in the differential equation:

$$mL^2\ddot{\varphi} + ca^2\varphi = 0; \quad mL^2\ddot{\varphi} + a^2c\varphi = 0$$

$$\ddot{\varphi} + \frac{a^2c}{mL^2} = 0$$

2.5. The rope is coiled on a homogeneous cylinder. The mass of the cylinder is *M*, radius *R*, and it rotates without friction around axis *O*, The weight in mass *m* is hooked up at the other end of the rope (Figure 2.8.). The spring with rigidity *c* is placed at distance $a$ from axis *O*. The other end of the spring is fixed for an immovable wall. Write a differential equation, calculate circular frequency $\omega$ and period *T* using the law of change of angular momentum. It is known: $c$=1560 [N/m]; $a$=0,2 [m]; $R$=0,35 [m]; $m$=5,2 [kg]; $M$=11,7 [kg].

Solution:

The law of change of angular momentum for the axis in point $O$, which is normal with plane of the sketch (Figure 2.8.):

$$J_O \cdot \ddot{\varphi} = \sum M_O^{Fi},$$

where $J_O$ is the moment of inertia from vibration's system for axis in point $O$, and $M_O^{Fi}$ is the moment of force $F_i$ for axis in point $O$.

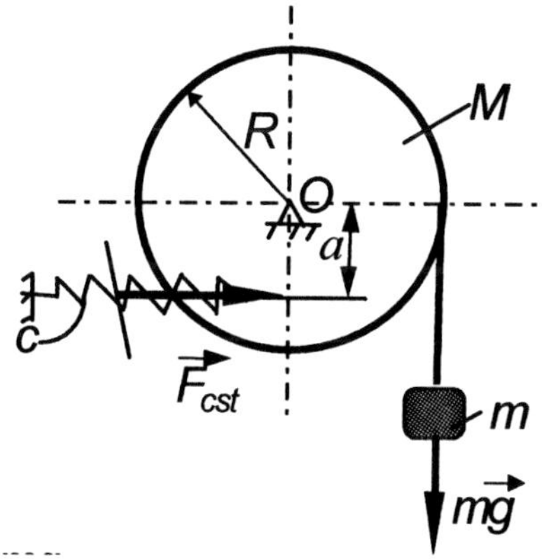


*Figure 2.8.*

Displace the system off the position of equilibrium (Figure 2.9.). <u>If you rotate the system clockwise, the force's moment is positive in the same direction of rotation.</u>

$$J_o \cdot \ddot{\varphi} = \sum M_o^{Fi} \Leftrightarrow J_o \cdot \ddot{\varphi} = mgR - F_c a \cdot \cos\varphi$$

Moment of inertia:

$$J_o = \frac{1}{2} MR^2 + 0 + mR^2$$

$\cos\varphi$ is approximated with 1 for small vibrations (when the angle $\varphi$ is small)

$$\left(\frac{1}{2} MR^2 + mR^2\right)\ddot{\varphi} = mgR - c \cdot (f_{st} + a\varphi) \cdot a$$

$$\left(\frac{1}{2} MR^2 + mR^2\right)\ddot{\varphi} = mgR - c \cdot f_{st} \cdot a - ca^2\varphi \quad .... \text{(2.2.6.)}$$

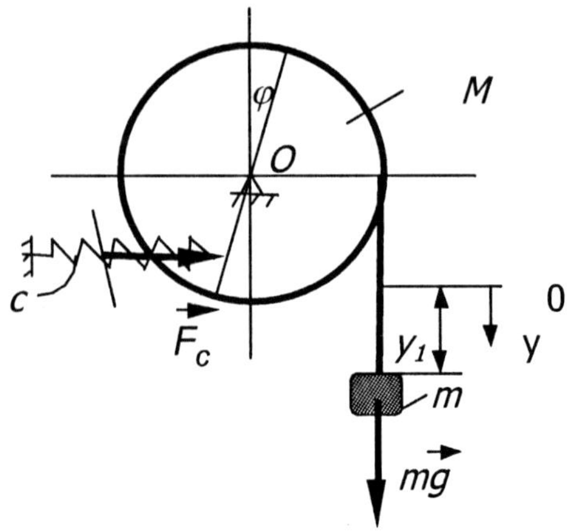


*Figure 2.9.*

Static condition of equilibrium (Figure 2.8.):

$$\sum M_o = 0 \Leftrightarrow F_{cst} a - mgR = 0$$

$$cf_{st} a - mgR = 0$$

When we put it in the differential equation (2.2.6.):

$$\left(\frac{1}{2}MR^2 + mR^2\right)\ddot{\varphi} = -ca^2\varphi \quad \Leftrightarrow \quad \ddot{\varphi} + \frac{ca^2}{R^2\left(m + \dfrac{M}{2}\right)}\varphi = 0$$

The circular frequency:

$$\omega = \frac{a}{R}\sqrt{\frac{c}{m + \dfrac{M}{2}}} = \frac{0,2}{0,35}\sqrt{\frac{1560\dfrac{N}{m}}{5,2kg + \dfrac{11,7kg}{2}}} = 6,79\left[\frac{rad}{s}\right]$$

The period:

$$T = \frac{2\pi}{\omega} = \frac{2\pi}{6,79\dfrac{rad}{s}} = 0,93[s]$$

2.6. Solve the last problem using LaGrange's equation.

Solution:

The kinetic energy of vibration system:

$$E_k = \frac{1}{2} m V_m^2 + \frac{1}{2} J_o \dot{\varphi}^2$$

$$E_k = \frac{1}{2} m R^2 \dot{\varphi}^2 + \frac{1}{2}\left(\frac{1}{2} M R^2\right)\dot{\varphi}^2$$

$$\frac{\partial E_k}{\partial \dot{\varphi}} = 2 \cdot \frac{1}{2} m R^2 \dot{\varphi} + 2 \cdot \frac{1}{2}\left(\frac{1}{2} M R^2\right)\dot{\varphi}, \quad \frac{\partial E_k}{\partial \varphi} = 0$$

Potential energy in Figure 2.9. ($\sin\varphi$ is approximated with $\varphi$ for small angle $\varphi$):

$$E_p = -mgy_1 + \frac{1}{2} c (f_{st} + a\varphi)^2, \qquad y_1 = R\varphi$$

$$E_p = -mgR\varphi + \frac{1}{2} c (f_{st} + a\varphi)^2$$

$$\frac{\partial E_p}{\partial \varphi} = -mgR + 2 \cdot \frac{1}{2} c (f_{st} + a\varphi) a$$

Equation of LaGrange:

$$\frac{d}{dt}\left(\frac{\partial E_k}{\partial \dot{\varphi}}\right) - \frac{\partial E_k}{\partial \varphi} + \frac{\partial E_p}{\partial \varphi} = 0$$

$$2\frac{1}{2} m R^2 \ddot{\varphi} + 2\frac{1}{4} M R^2 \ddot{\varphi} - mgR + 2\frac{1}{2} c (f_{st} + a\varphi) a = 0$$

$$\left(mR^2 + \frac{1}{2} M R^2\right)\ddot{\varphi} - mgR + caf_{st} + ca^2\varphi = 0 \quad \ldots . (2.2.7.)$$

The static condition of equilibrium (Figure 2.8.):

$$\sum M_o = 0 \Leftrightarrow F_{cst} a - mgR = 0$$

$$cf_{st} a - mgR = 0$$

Put it in the differential equation (2.2.7.):

$$\left(\frac{1}{2} M R^2 + m R^2\right) \cdot \ddot{\varphi} + ca^2 \varphi = 0$$

$$\ddot{\varphi} + \frac{ca^2}{R^2\left(m + \frac{M}{2}\right)}\varphi = 0$$

2.7.The homogeneous disc, with mass $m$ and radius $r$, does small vibrations inside the cylindrical half-roll, with radius $R$=11$r$. The cylindrical spring with rigidity c is fixed for axis of disc. How much would the rigidity of spring be if the circular frequency was $\omega=\sqrt{10}$ [rad/s]? It is given: $m$=20 [kg], $r$=0,981 [m].

Solution:

The kinetic energy from vibration's system:

$$E_k = \frac{1}{2}mv_T^2 + \frac{1}{2}\left(\frac{1}{2}mr^2\right)\dot{\theta}^2$$

$$E_k = \frac{1}{2}m(R-r)^2\dot{\varphi}^2 + \frac{1}{4}mr^2\dot{\theta}^2$$

We can see that arcs $P_vA$ and $P_vB$ are equal in Figure 2.11.

$$(\theta+\varphi)r = R\varphi$$

$$\theta = \frac{R-r}{r}\varphi$$

Let us derive the last equation with time $t$:

$$\frac{d\theta}{dt} = \frac{R-r}{r}\frac{\varphi}{dt}$$

$$\dot{\theta} = \frac{R-r}{r}\dot{\varphi}$$

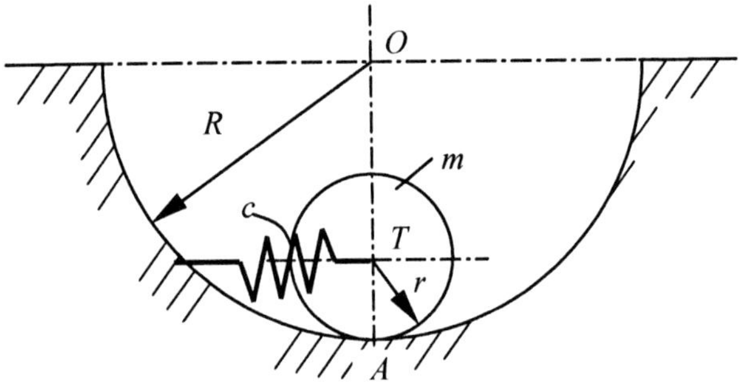


*Figure 2.10.*

Now the kinetic energy:

$$E_k = \frac{1}{2}m(R-r)^2\dot{\varphi}^2 + \frac{1}{4}m(R-r)^2\dot{\varphi}^2$$

$$E_k = 75mr^2\dot{\varphi}^2$$

$$E_p = mgh + \frac{1}{2}c(10r\sin\varphi)^2$$

The height *h* is defined in Figure 2.11.:

$$h = 10r - 10r\cos\varphi = 10r(1-\cos\varphi)$$

For small vibrations, **when we use LaGrange's equation** for solving the problem, it is necessary to approximate $\cos\varphi = 1 - \frac{\varphi^2}{2}$ and $\sin\varphi$ with $\varphi$:

$$h \approx 10r\left[1 - \left(1 - \frac{\varphi^2}{2}\right)\right] = 10r\frac{\varphi^2}{2} = 5r\varphi^2$$

$$E_p = mg \cdot 5r\varphi^2 + \frac{1}{2}c(10r\varphi)^2$$

$$E_p = 5mgr\varphi^2 + 50cr^2\varphi^2$$

LaGrange's equation:

$$\frac{d}{dt}\left(\frac{\partial E_k}{\partial \dot{\varphi}}\right) - \frac{\partial E_k}{\partial \varphi} + \frac{\partial E_p}{\partial \varphi} = 0$$

$$75 \cdot mr^2 \cdot 2 \cdot \ddot{\varphi} + 2 \cdot 5mgr\varphi + 2 \cdot 50cr^2\varphi = 0$$

$$150mr^2\ddot{\varphi} + \left(10mgr + 100cr^2\right) \cdot \varphi = 0$$

$$\ddot{\varphi} + \frac{mg + 10cr}{15mr} \cdot \varphi = 0$$

$$\omega = \sqrt{\frac{mg + 10cr}{15mr}} \Rightarrow \quad \omega^2 \cdot 15mr = mg + 10cr$$

$$10cr = \omega^2 \cdot 15mr - mg$$

$$c = \frac{15}{10}\omega^2 m - \frac{mg}{10r}$$

$$c=\frac{15}{10}\cdot 10\frac{rad^2}{s^2}\cdot 20kg-\frac{20kg\cdot 9{,}81\dfrac{m}{s^2}}{10\cdot 0{,}981m}=280\left[\frac{N}{m}\right]$$

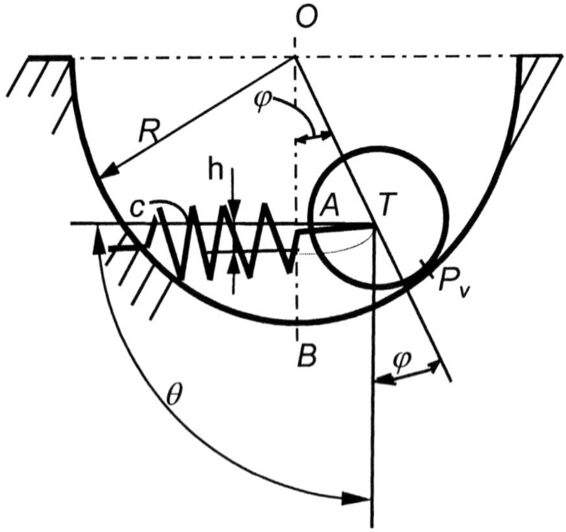


Figure 2.11.

2.8. The concentrated mass *m* is placed at the end of the double lever. The mass of the lever is neglected. The rope is hitched at the other end of the lever and coiled round the homogeneous disc. The disc has mass *M*, radius *R* and it rotates round axis $O_2$ without friction. The torsion spring, with rigidity *c*, is placed on distance $a$ from the axis $O_2$. Write the differential equation of small vibrations, the circular frequency, and the time of swing.
It is known: *c*=2500[N/m], $a$=0,25[m], *R*=40[cm], *m*=2,1[kg], *M*=4,2[kg].

Solution:
$E_{km}$ - the kinetic energy of mass *m*, $E_{kM}$ - the kinetic energy of mass *M*

$$E_k=E_{km}+E_{kM}$$

$$E_{kM}=\frac{1}{2}m\cdot v_m^2=\frac{1}{2}m\left(l\cdot\dot{\theta}\right)^2$$

$$E_{kM}=\frac{1}{2}J_{O2}\cdot\dot{\varphi}^2$$

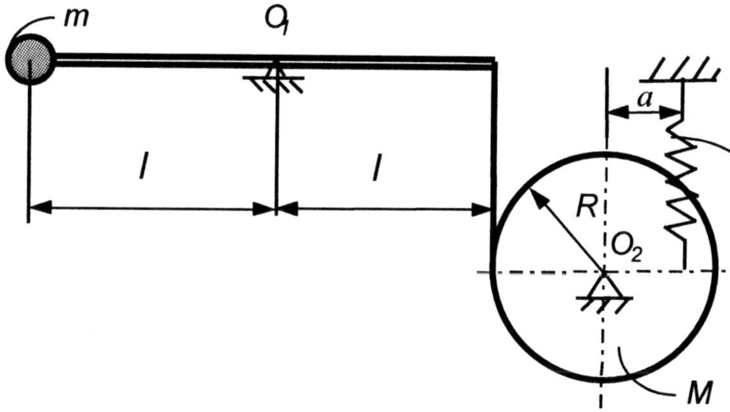


*Figure 2.12.*

The relation between the angles $\theta$ and $\varphi$ (Figure 2.13.):

$$l\theta=R\varphi \quad ; \quad \theta=\frac{R}{l}\varphi$$

If we derive the last equation with time *t*, we will get:

$$\dot{\theta}=\frac{R}{l}\dot{\varphi}$$

$$E_k=\frac{1}{2}ml^2\frac{R^2}{l^2}\dot{\varphi}^2=\frac{1}{2}mR^2\dot{\varphi}^2+\frac{1}{4}MR^2\dot{\varphi}^2$$

The potential energy of the system:

$$E_p=E_{pm}+E_{pc}$$

where: $E_{pm}$ is the potential energy of mass *m* and $E_{pc}$ is the potential energy of the spring.

$$E_{pm}=-mgh=-mgl\sin\theta\approx-mgl\theta=-mgl\frac{R}{l}\varphi$$

$$E_{pc}=\frac{1}{2}c(f_{st}+a\varphi)^2$$

$$E_p=-mgR\varphi+\frac{1}{2}c(f_{st}+a\varphi)^2$$

LaGrange's equation:

$$\frac{d}{dt}\left(\frac{\partial E_k}{\partial \dot{\varphi}}\right)-\frac{\partial E_k}{\partial \varphi}+\frac{\partial E_p}{\partial \varphi}=0$$

$$2\cdot\frac{1}{2}mR^2\ddot{\varphi}+2\cdot\frac{1}{4}MR^2\ddot{\varphi}-mgR+2\cdot\frac{1}{2}c\left(f_{st}+a\varphi\right)a=0$$

$$\left(mR^2+\frac{1}{2}MR^2\right)\ddot{\varphi}-mgR+caf_{st}+ca^2\varphi=0$$

We get the static condition of equilibrium from Figure 2.14, where the system is shown with the rope's cross section. The internal rope's force is *S*:

$$\sum M_{O1}=0$$

$$mgl-Sl=0$$

$$S=mg$$

$$\sum M_{O2}=0$$

$$SR-F_{cst}a=0$$

$$mgR-cf_{st}a=0$$

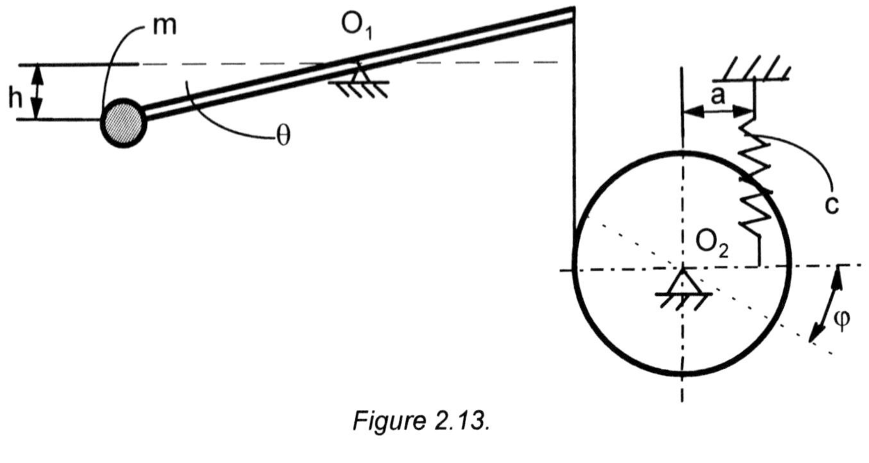


*Figure 2.13.*

When we put the static condition of equilibrium in the differential equation, we will get:

$$\left(mR^2+\frac{1}{2}MR^2\right)\ddot{\varphi}+ca^2\varphi=0$$

$$\ddot{\varphi} + \frac{ca^2}{mR^2 + \frac{1}{2}MR^2}\varphi = 0$$

$$\omega = \sqrt{\frac{ca^2}{mR^2 + \frac{1}{2}MR^2}} = \sqrt{\frac{2500\left[\frac{N}{m}\right]\cdot 0{,}25^2[m]^2}{2{,}1[kg]\cdot(0{,}4[m])^2 + \frac{1}{2}4{,}2[kg]\cdot(0{,}4[m])^2}} =$$

$$= 15{,}25\left[\frac{rad}{s}\right]$$

$$T = \frac{2\pi}{\omega} = \frac{2\pi}{15{,}25\left[\frac{rad}{s}\right]} = 0{,}41[s]$$

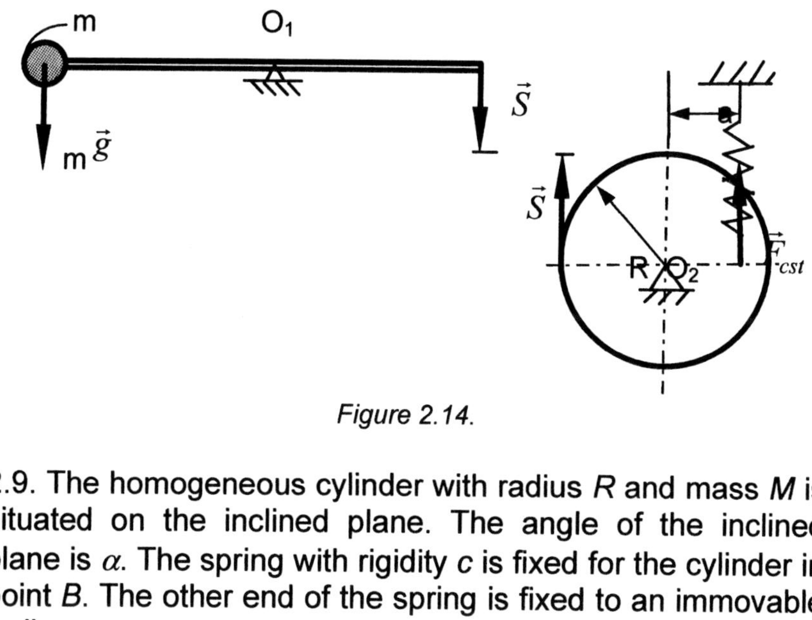


*Figure 2.14.*

2.9. The homogeneous cylinder with radius *R* and mass *M* is situated on the inclined plane. The angle of the inclined plane is $\alpha$. The spring with rigidity *c* is fixed for the cylinder in point *B*. The other end of the spring is fixed to an immovable wall.

Estimate the period of rolling vibrations. (The cylinder rolls without sliding.)

It is known: *c*=1000 [N/m]; *M*=16,8 [kg]

Solution:

The static condition of equilibrium (Figure 2.15.):

$$\sum M_{Pv} = 0$$

$$F_{cst} \cdot 2R - MgR\sin\alpha = 0$$

$$cf_{st} \cdot 2R - MgR\sin\alpha = 0$$

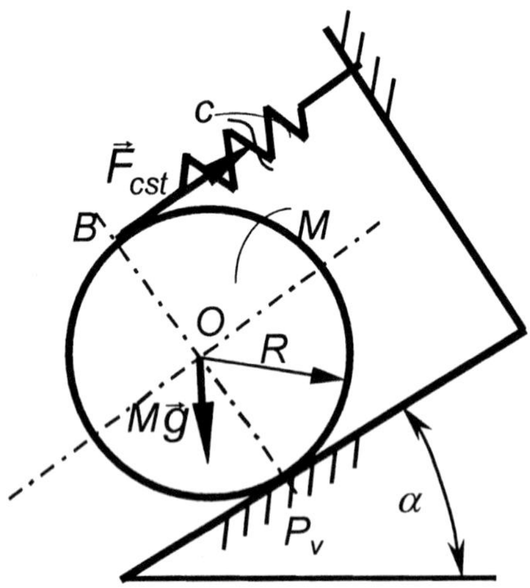


*Figure 2.15.*

The cylinder does the plane movement and the kinetic energy is composed of rotary ($E_{krot}$) and translatory ($E_{ktr}$) component:

$$E_k = E_{krot} + E_{ktr}$$

$$E_{krot} = \frac{1}{2} J_0 \cdot \dot{\varphi}^2 = \frac{1}{2}(\frac{1}{2} MR^2) \cdot \dot{\varphi}^2$$

$$E_{ktr} = \frac{1}{2} M \cdot v_0^{\,2} = \frac{1}{2} M \cdot (R \cdot \dot{\varphi})^2$$

$$E_k = \frac{1}{4} MR^2 \cdot \dot{\varphi}^2 + \frac{1}{2} MR^2 \cdot \dot{\varphi}^2$$

$$E_p = E_{pc} + E_{pM}$$

$$E_{pc} = \frac{1}{2} c \cdot (f_{st} + 2R\varphi)^2$$

$$E_{pM} = -Mgh = -M \cdot g \cdot (R\varphi \cdot \sin\alpha)$$

$$E_p = \frac{1}{2} c \cdot (f_{st} + 2r\varphi)^2 - MgR\varphi \sin\alpha$$

LaGrange's equation:

$$\frac{d}{dt}\left(\frac{\partial E_k}{\partial \dot{\varphi}}\right) - \frac{\partial E_k}{\partial \varphi} + \frac{\partial E_p}{\partial \varphi} = 0$$

$$2\frac{1}{4}MR^2\ddot{\varphi} + 2\frac{1}{2}MR^2\ddot{\varphi} + 2\frac{1}{2}c(f_{st} + 2R\varphi)2R - MgR\sin\alpha = 0$$

$$\left(\frac{1}{2}MR^2 + MR^2\right)\ddot{\varphi} + cf_{st} \cdot 2R + c \cdot 4R^2 \cdot \varphi - MgR\sin\alpha = 0$$

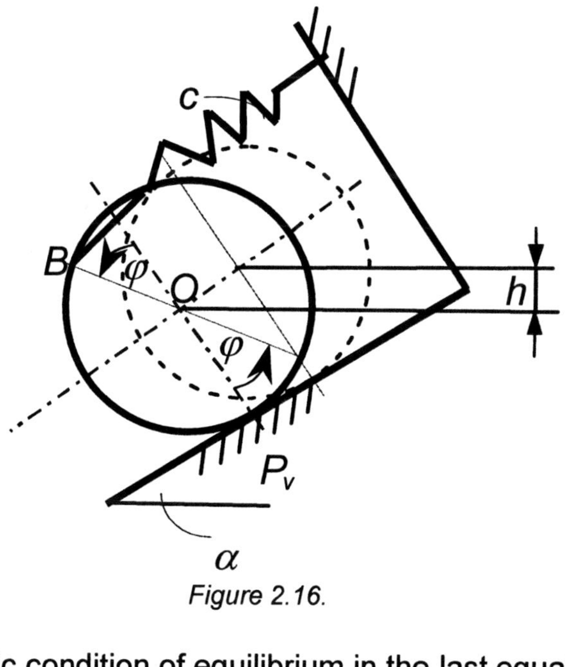


*Figure 2.16.*

Put the static condition of equilibrium in the last equation:

$$\frac{3}{2}MR^2 \cdot \ddot{\varphi} + c \cdot 4R^2 \cdot \varphi = 0$$

$$\ddot{\varphi} + \frac{4c \cdot R^2}{\frac{3}{2}MR^2} \cdot \varphi = 0$$

$$\ddot{\varphi} + \frac{8}{3} \cdot \frac{c}{M} \cdot \varphi = 0$$

$$T = \frac{2\pi}{\omega} = \frac{2\pi}{\sqrt{\frac{8}{3} \cdot \frac{c}{M}}} = \frac{2\pi}{\sqrt{\frac{8}{3} \cdot \frac{1000\left[\frac{N}{m}\right]}{16,8[kg]}}} = 0,5 \quad [s]$$

2.10. The homogeneous disc with mass *M* and radius *R* is situated on an inclined plane, the axis of the disc is linked with the spring (Figure 2.17.). The spring has rigidity *c*. The rope is coiled around the disc and the other end is thrown over the roller. Radius of the roller is *r*, and its mass is neglected. The weight with mass *m* is hooked up at the other end of the rope.
Estimate the circular frequency $\omega$ and the time of the swing *T*. Neglect all frictional resistances. It is given: *c*=2000 [N/m], *R*=0,3 [m], *m*=2 [kg], *M*=5 [kg].

Solution:

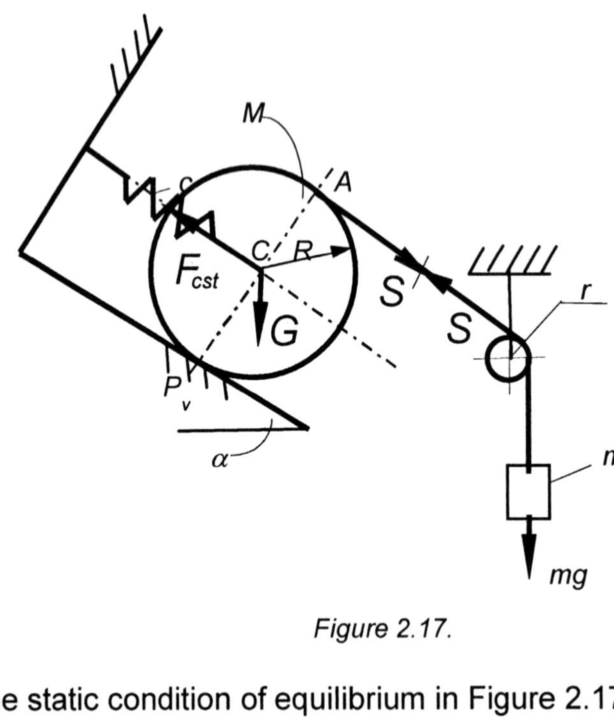


*Figure 2.17.*

The static condition of equilibrium in Figure 2.17:

$$\sum M_O = 0$$

$$mg \cdot r - S \cdot r = 0 \quad \Rightarrow \quad S = mg$$

$$\sum M_{pv} = 0$$

$$F_{cst} \cdot R - S \cdot 2R - MgR\sin\alpha = 0$$

$$c \cdot f_{st} - mg \cdot 2R - MgR\sin\alpha = 0$$

$$cf_{st} - 2mg - Mg\sin\alpha = 0$$

It is needed to estimate the kinetic and potential energy of the system:

$$E_k = E_{km} + E_{kM}$$

$$E_{km} = \frac{1}{2} m \cdot v_m^2 = \frac{1}{2} m \cdot v_A^2 = \frac{1}{2} m \cdot (2R\dot{\varphi})^2$$

$$E_{kM} = E_{kM}^{rot} + E_{kM}^{tr} = \frac{1}{2} J_c \cdot \dot{\varphi}^2 + \frac{1}{2} Mv_c^2 =$$

$$= \frac{1}{2}\left(\frac{1}{2} MR^2\right) \cdot \dot{\varphi}^2 + \frac{1}{2} M(R\dot{\varphi})^2$$

$$E_k = 2mR^2\dot{\varphi}^2 + \frac{1}{4} MR^2\dot{\varphi}^2 + \frac{1}{2} MR^2\dot{\varphi}^2$$

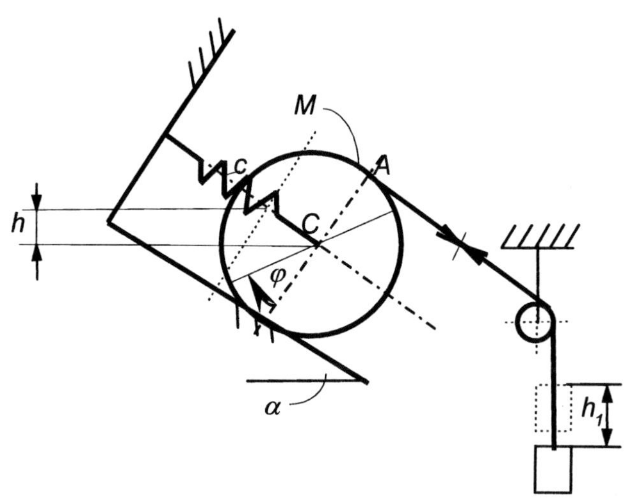


*Figure 2.18.*

$$E_p = E_{pm} + E_{pM} + E_{pc}$$

$$E_{pm} = -mgh_1 = -mg(2R\varphi)$$

$$E_{pM} = -Mgh_2 = -Mg(r\varphi \sin\alpha)$$

$$E_{pc} = \frac{1}{2}c \cdot (f_{st} + R\varphi)^2$$

$$E_p = -2mgR\varphi - MgR\varphi\sin\alpha + \frac{1}{2}c \cdot (f_{st} + R\varphi)^2$$

LaGrange's equation:

$$\frac{d}{dt}\left(\frac{\partial E_k}{\partial \dot{\varphi}}\right) - \frac{\partial E_k}{\partial \varphi} + \frac{\partial E_p}{\partial \varphi} = 0$$

$$4mR^2 \cdot \ddot{\varphi} + 2 \cdot \frac{1}{4}MR^2 \cdot \ddot{\varphi} + 2 \cdot \frac{1}{2}MR^2 \cdot \ddot{\varphi} - 2mgR -$$

$$- MgR\sin\alpha + 2 \cdot \frac{1}{2} \cdot c\left(f_{st} + R \cdot \varphi\right) \cdot R = 0$$

$$(4mR + \frac{1}{2}MR + MR^2) \cdot \ddot{\varphi} - 2mg - Mg\sin\alpha + cf_{st} + cR\varphi = 0$$

Put the static condition of equilibrium:

$$\left(4mR + \frac{3}{2}MR\right) \cdot \ddot{\varphi} + cR\varphi = 0 \ ; \qquad \ddot{\varphi} + \frac{cR}{4mR + \frac{3}{2}MR} \cdot \varphi = 0$$

$$\omega^2 = \frac{cR}{4mR + \frac{3}{2}MR} = \frac{2000\left[\frac{N}{m}\right] \cdot 0{,}3[m]}{4 \cdot 2[kg] \cdot 0{,}3[m] + \frac{3}{2}5[kg] \cdot 0{,}3[m]} = 129{,}0\left[s^{-2}\right]$$

$$\omega = 11{,}36\left[s^{-1}\right]$$

$$T = \frac{2\pi}{\omega} = 0{,}55[s]$$

2.11. Unextensible rope is linked for immovable ground with one end and thrown over the disc (Figure 2.19.). The disc has mass *M* and radius *R*. The other end of the rope is linked to a rigid lever. The length of the lever is 2*l*, its weight is *G* and it rotates around joint O. The disc's axis is linked to the ceiling with a spring. The rigidity of the spring is *c*. All frictions are neglected. Calculate $\omega$ and $T$, if it is known: $G$, $M$, $c$ and $R=l$.

Solution:
The kinetic energy of vibration's system is composed of kinetic energy of the disc $E_{kM}$ and the lever $E_{kG}$:

$$E_k = E_{kG} + E_{kM}$$

$$E_{kG} = \frac{1}{2} J_0 \cdot \varphi^2 = \frac{1}{2}\left[\frac{1}{3}\frac{G}{g}(2l)^2\right]\dot{\varphi}^2$$

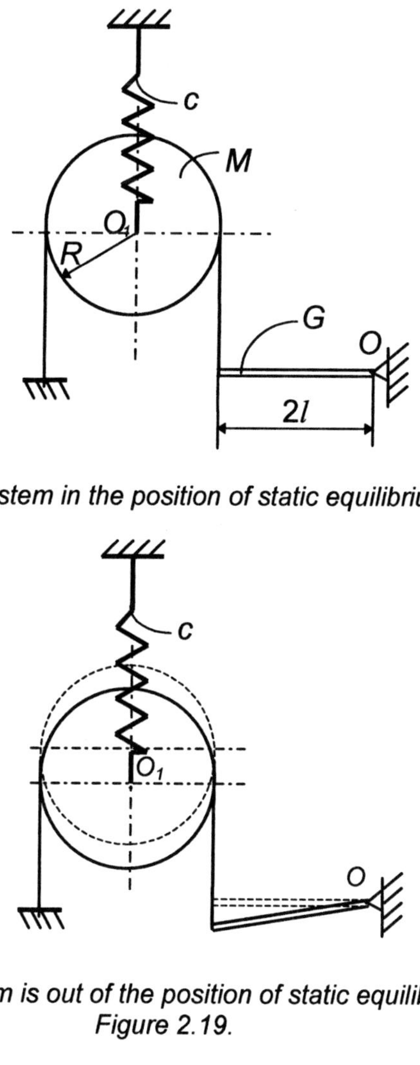


a) The system in the position of static equilibrium

b) The system is out of the position of static equilibrium
Figure 2.19.

Where $J_O$ is disc's moment of inertia

$$E_{kM} = E_{kM}^{tr} + E_{kM}^{rot} = \frac{1}{2} M v_{O1}^2 + \frac{1}{2} J_{O1} \dot{\theta}^2$$

$$EkM = \frac{1}{2} M\left(R\dot{\theta}\right)^2 + \frac{1}{2}\left(\frac{1}{2} MR^2\right)\dot{\theta}^2$$

Relation between the angles $\theta$ and $\varphi$ in Figure 2.20:

$$2R\theta = 2l\varphi \quad \Leftrightarrow \quad 2R\theta = 2R\varphi \quad \Rightarrow \quad \theta = \varphi$$

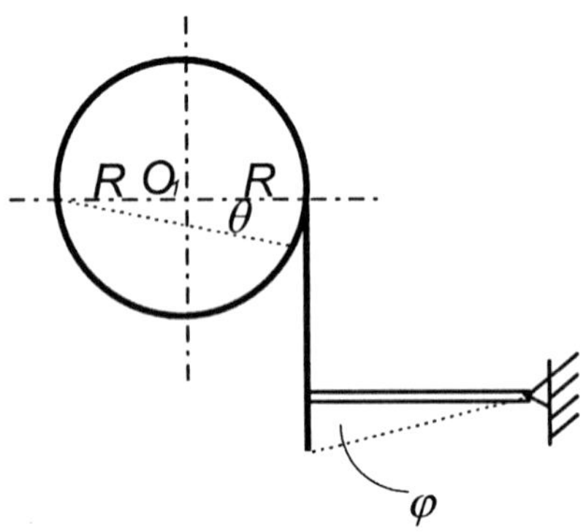


*Figure 2.20.*

Kinetic energy of the system:

$$E_k = \frac{2}{3}\frac{G}{g} l^2 \dot{\varphi}^2 + \frac{1}{2} MR^2 \dot{\varphi}^2 + \frac{1}{4} MR^2 \dot{\varphi}^2$$

$$E_k = \frac{2}{3}\frac{G}{g} l^2 \dot{\varphi}^2 + \frac{3}{4} MR^2 \dot{\varphi}^2$$

The potential energy:

$$E_p = E_{pG} + E_{pM} + E_{pc}$$

$$E_{pG} = -Gl\sin\varphi \approx -Gl\varphi$$

$$E_{pM} = -R\theta\, Mg = -MgR\varphi$$

$$E_{pc} = \frac{1}{2} c\left(f_{st} + R\theta\right)^2 = \frac{1}{2} c\left(f_{st} + R\varphi\right)^2$$

$$E_p = -Gl\varphi - MgR\varphi + \frac{1}{2} c\left(f_{st} + R\varphi\right)^2$$

LaGrange's equation:

$$\frac{d}{dt}\left(\frac{\partial E_k}{\partial \dot{q}}\right)-\frac{\partial E_k}{\partial q}+\frac{\partial E_p}{\partial q}=0$$

$$2\frac{2}{3}\frac{G}{g}l^2\ddot{\varphi}+2\frac{3}{4}MR^2\ddot{\varphi}+(-Gl)-MgR+2\frac{1}{2}c\left(f_{st}+R\varphi\right)R=0$$

$$\frac{4}{3}\frac{G}{g}l^2\ddot{\varphi}+\frac{3}{2}Ml^2\ddot{\varphi}-Gl-Mgl+clf_{st}+cl^2\varphi=0$$

The static condition of equilibrium in Figure 2.21:

$$\sum M_O=0$$

$$G\cdot l-S\cdot 2l=0 \quad\Rightarrow\quad S=\frac{G}{2}$$

$$\sum M_{pv}=0$$

$$F_{cst}\cdot R-Mg\cdot R-S\cdot 2\cdot R=0 \quad\Leftrightarrow\quad cf_{st}\cdot R-MgR-\frac{G}{2}\cdot 2R=0$$

$$cf_{st}\cdot l-Mgl-G\cdot l=0$$

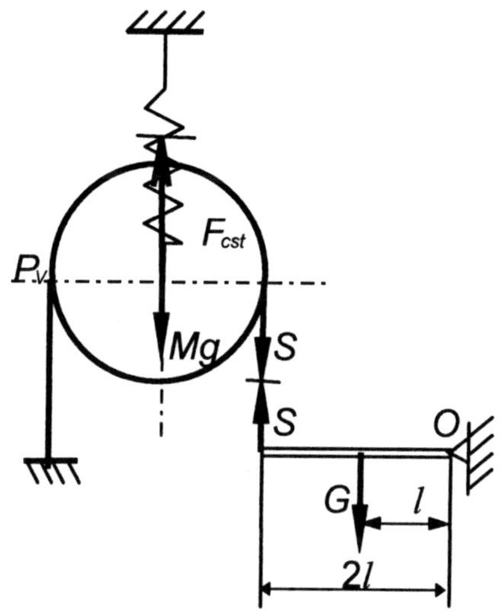


*Figure 2.21.*

Put the static condition in the differential equation:

$$\left(\frac{4}{3}\frac{G}{g}l^2+\frac{3}{2}Ml^2\right)\cdot\ddot{\varphi}+cl^2\varphi=0$$

$$\left(\frac{4}{3}\frac{G}{g}+\frac{3}{2}M\right)\cdot\ddot{\varphi}+c\varphi=0$$

$$\ddot{\varphi}+\frac{c}{\frac{4}{3}\frac{G}{g}+\frac{3}{2}M}\cdot\varphi=0$$

$$\omega=\sqrt{\frac{c}{\frac{4}{3}\frac{G}{g}+\frac{3}{2}M}}$$

$$T=\frac{2\pi}{\omega}=2\pi\sqrt{\frac{\frac{4}{3}\frac{G}{g}+\frac{3}{2}M}{c}}$$

2.12. Vibration's system is shown in Figure 2.22. in the position of equilibrium. Calculate the circular frequency $\omega$, the time of swing $T$ and the law of vibrations. It is known: $M$=5[kg], $R$=0,3[m], $c$=2000[N/m] and initial conditions: $t_0$=0[s], $\varphi_0$=0[rad], $\dot{\varphi}_0$=0,1[m/s].

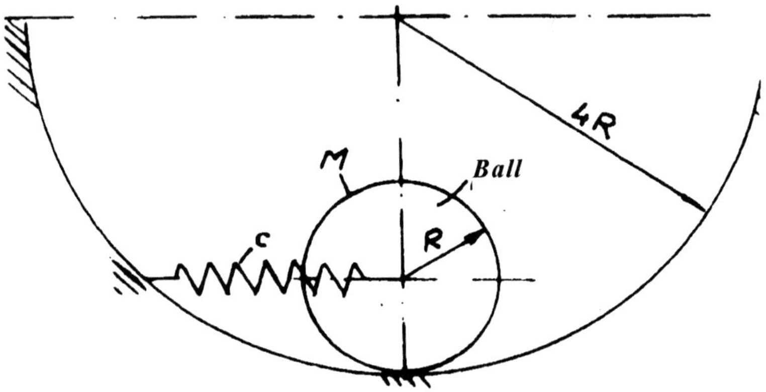


*Figure 2.22.*

(Hint: The sphere's moment of inertia for axis in the centre is $J_o = \frac{2}{5} \cdot MR^2$, see table 7. in appendix.)

Final solution:

$\omega$=17,13[rad/s], $T$=0,367[s],

$\varphi = 0{,}005 \cdot \sin 17{,}13t$ [m,s]

2.13. Vibration's system is shown in Figure 2.23. in the position of equilibrium. Estimate the circular frequency $\omega$, the time of swing $T$ and the law of vibrations, if It is given: $M$=7[kg], $l$=0,5[m], $c_1$=1800[N/m], $c_2$=1500[N/m]. The initial conditions are: $t_0$=0[s], $\varphi_0$=0[rad] and $\dot{\varphi}_0$=0,1[m/s].

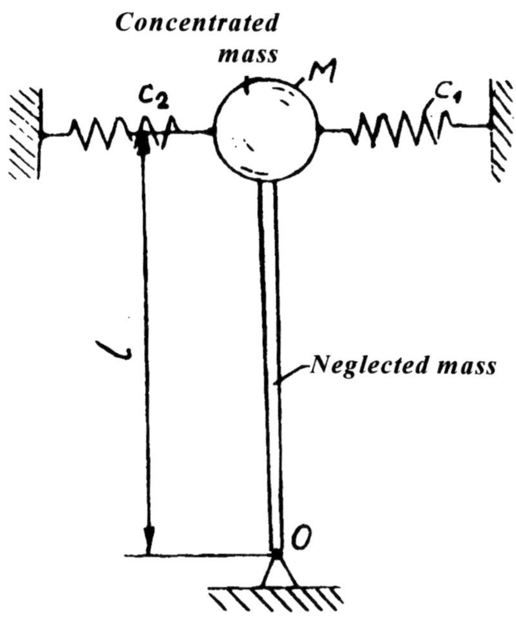


Figure 2.23.

Final solution:

$\omega$=21,26[rad/s], T=0,296[s],

$\varphi = 0{,}005 \cdot \sin 21{,}26t$ [m,s]

2.14. Vibration's system is shown in Figure 2.24. in position of equilibrium. Estimate the circular frequency $\omega$, the time of swing $T$ and the law of vibrations, if It is known: $a$=0,5[m]; m=7[kg]; $c_1$=1500[N/m]; $c_2$=2000[N/m]; the inertia's moment of lever $AOB$ for axis in point $O$: $J_o$=0,7[kgm$^2$]; b=0,3[m]. Lever's centre of gravity is in point $O$. The initial conditions are: $t_0$=0[s]; $\varphi_0$=0[rad] and $\dot{\varphi}_0$=0,1[m/s].

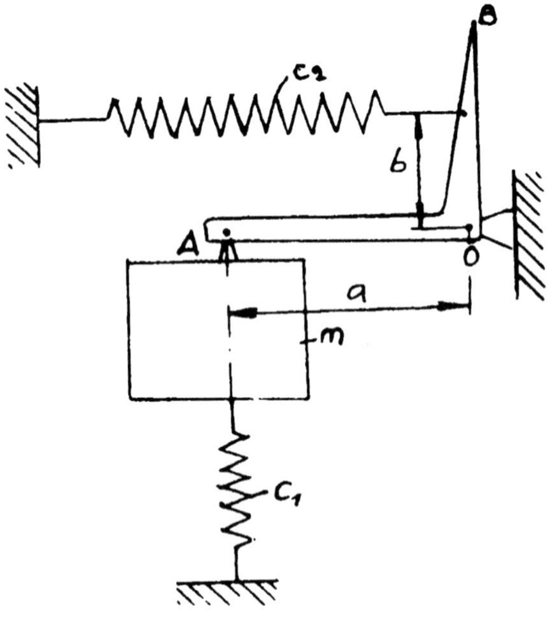


*Figure 2.24.*

Final solution:
$\omega$=15,05[rad/s], T=0,417[s],
$\varphi = 0{,}007 \cdot \sin 15{,}05t$ [m,s]

# CHAPTER THREE

## APPLICATION OF LAGRANGE - DIRIHLET THEOREM ABOUT THE STABILITY OF THE CONSERVATIVE SYSTEM EQUILIBRIUM

### 3.1. Short Theory Review

LaGrange - Dirichlet's theorem about the stability of the conservative system's equilibrium:
"If the potential energy of vibration's system is minimal for some position, it is the position of stable equilibrium."
There are two kinds of problems to address here:
3.1.1. It is given the position of equilibrium in the problem and that is needed to establish the condition that the position would be stable.
It is necessary to estimate the potential energy of the system ($E_p$) and the second derivative of $E_p$ for a generalized coordinate $\left(\frac{\partial^2 E_p}{\partial q^2}\right)$, than to insert a value of generalized coordinate for the position of equilibrium in the derivative and it must be bigger than 0:

$$\left(\frac{\partial^2 E_p}{\partial q^2}\right)_{q=q_o} > 0 \qquad \dots\dots\dots\dots\dots (3.1.)$$

To avoid complicated trigonometric equations, it is possible to approximate (for example if we have an angle $\alpha$) $\sin\alpha \approx \alpha;\ \cos\alpha \approx 1-\frac{\alpha^2}{2}$
3.1.2. It isn't given the position of equilibrium in the problem and it is needed to define potential energy ($E_p$) like the function of generalized coordinate ($q$). We mustn't approximate trigonometric expressions in the problems. Then it's necessary to estimate the first partial derivative of $E_p$ for generalized coordinate, it equals zero:

$$\frac{\partial E_p}{\partial q} = 0 \quad \ldots\ldots\ldots\ldots\ldots\ldots\ldots \quad (3.2)$$

Solutions of equation (3.2.) (it is possible that we have more than one solution) are the positions of vibration's system equilibrium. Furthermore, it is needed to find the conditions of stability for each position of equilibrium. The procedure is described in 3.1.1. (inequality (3.1.)).

## 3.2. Problems

3.1. Unextensible rope is linked for immovable ground with the end and thrown over the disc. The disc has mass *M* and radius *R=l*. The other end of the rope is linked to a rigid lever. The length of the lever is *2l*, its weight is G, and it rotates around joint O. The axis of the disc is linked to the ceiling with the spring. Rigidity of the spring is *c*. All frictions are neglected. Estimate the condition of stability for the position of equilibrium shown in Figure 3.1.

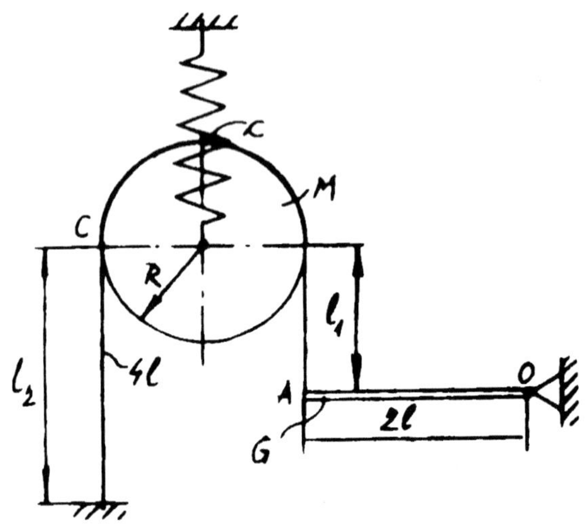


*Figure 3.1.*

Solution:

Relation of angles $\varphi$ and $\theta$ in Figure 3.3:

$2l\varphi = 2R\theta$

$\varphi = \theta$

Potential energy of the system ($E_p$) is composed of: potential energy of the lever ($E_{pl}$), the disc ($E_{pM}$) and the spring ($E_{pc}$):

$$E_p = E_{pl} + E_{pM} + E_{pc}$$

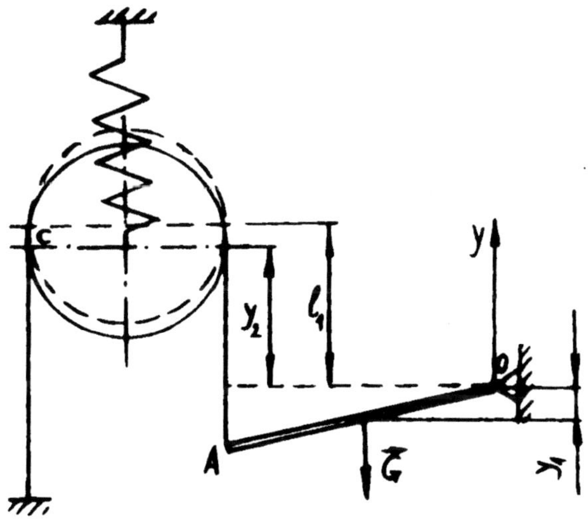


*Figure 3.2.*

The method of coordinates is the most favorable for the definition of potential energy in applying LaGrange's - Dirihlet theorem. It is necessary to displace the system off the position of equilibrium (Figure 3.2.), then to select a point for origin for coordinates (point $O$ in figure 3.2.). After that it is necessary to define a vertical coordinate for each center of gravity in vibration's system.

In this problem we have two masses: the lever and the disc. $y_1$ is a vertical coordinate of the lever's center of gravity and $y_2$ of the disc in Figure 3.2.

Potential energy of the lever:

$$E_l = -Gy_1 = -Gl\sin\varphi \approx -Gl\varphi$$

Potential energy of the mass:

$$E_{pM} = Mgy_2 = Mg(l_1 - R\sin\theta) \approx Mg(l_1 - R\theta)$$

Potential energy of the spring:

$$E_{pc}=\frac{1}{2}c\left(f_{st}+R\sin\theta\right)^2\approx\frac{1}{2}c\left(f_{st}+R\theta\right)^2=\frac{1}{2}c\left(f_{st}+l\varphi\right)^2$$

Total potential energy:

$$E_p=-Gl\cdot\varphi+Mg\left(l_1-l\varphi\right)+\frac{1}{2}c\left(f_{st}+l\varphi\right)^2$$

$$\frac{\partial E_p}{\partial\varphi}=-Gl-Mgl+\frac{1}{2}c2\left(f_{st}+l\varphi\right)l=0$$

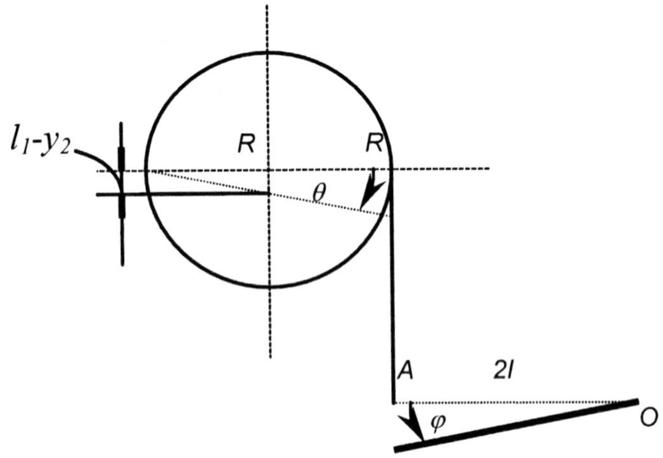


*Figure 3.3.*

Static condition of equilibrium (Figure 2.21. and 3.1.):

$$\sum M_o=0$$

$$Gl-S2l=0\Rightarrow S=\frac{G}{2}$$

$$\sum M_c=0$$

$$F_{cst}R-MgR-S2R=0$$

$$cf_{st}l-Mgl-Gl=0$$

Put in the first partial derivative $\frac{\partial E_p}{\partial\varphi}$:

$$cl^2\cdot\varphi=0\Rightarrow\varphi=0$$

Solution of equation $\frac{\partial E_p}{\partial \varphi}=0$ is $\varphi$=0, and it's position of equilibrium. Condition of stability (2.1.):

$$\left(\frac{\partial^2 E_p}{\partial \varphi^2}\right)_{\varphi=0} > 0$$

$$\frac{\partial^2 E_p}{\partial \varphi^2} = cl^2$$

$$cl^2 > 0$$

Physical values of rigidity *c* and length *l* are bigger than 0. So, the condition is surely accomplished, and the position in Figure 3.1. is stable.

3.2.Vibration's system is shown in the position of equilibrium in Figure 3.4. The radius of the homogeneous disc is *R* and length of the lever is *l*. Masses of the lever and the slider are *m*. Define the condition of the position's stability in the Figure. All resistance of movement is neglected.

Solution:

Relation of angles $\theta$ and $\varphi$ (Figure 3.5.):

$$l \cdot \sin\theta = R \cdot \sin\varphi$$

$$l\theta = l\varphi$$

$$\theta = \frac{R}{l}\varphi$$

Potential energy of the slider ($E_{pkl}$):

$$E_{pkl} = mgy_1 = mg(l\cos\theta - R\cos\varphi) = mg\left[l\left(1-\frac{\theta^2}{2}\right) - R\left(1-\frac{\varphi^2}{2}\right)\right] =$$

$$= mg\left(l - \frac{l\theta^2}{2} - R + \frac{R\varphi^2}{2}\right) = mg\left(l - \frac{R^2}{2l}\varphi^2 - R + \frac{R\varphi^2}{2}\right)$$

Potential energy of the lever ($E_{pst}$):

$$E_{pst} = mgy_2 = mg\left(y_1 - \frac{l}{2}\cos\theta\right) = mg\left(l\cos\theta - R\cos\varphi - \frac{l}{2}\cos\theta\right) =$$

$$= mg\left(\frac{l}{2}\cos\theta - R\cos\varphi\right) = mg\left[\frac{l}{2}\left(1-\frac{\theta^2}{2}\right) - R\left(1-\frac{\varphi^2}{2}\right)\right] =$$

$$= E_{pst} = mg\left(\frac{l}{2} - (R^2/4l)\varphi^2 - R + \frac{R}{2}\varphi^2\right)$$

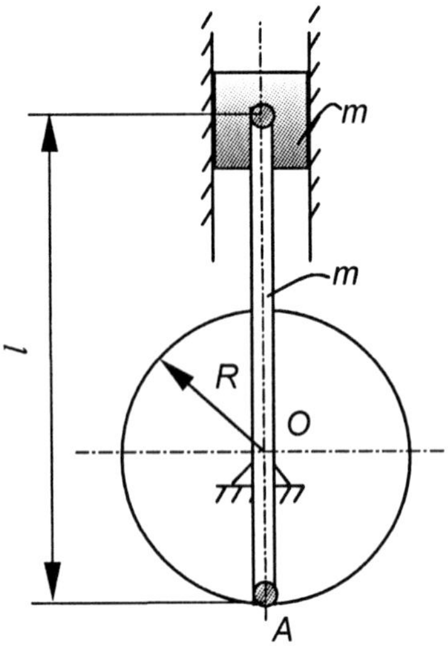


*Figure 3.4.*

Total potential energy ($E_p$):

$$E_p = E_{pkl} + E_{pst} = mgl - mg\frac{R^2}{2l}\varphi^2 - mgR + mgR\varphi^2/2 + mg\frac{l}{2} -$$

$$- mg\frac{R^2}{4l}\varphi^2 - mgR + mg\frac{R}{2}\varphi^2$$

After mathematical manipulation the last expression can be given as follows:

$$E_p = \frac{3mgl}{2} - 2mgR + \varphi^2\left(mgR - \frac{3mgR^2}{4l}\right)$$

$$\frac{\partial\ E_p}{\partial\varphi} = \left(mgR - \frac{3mgR^2}{4l}\right)2\varphi = 0$$

Solution of the last equation is $\varphi$=0. It's position of equilibrium is seen in Figure 3.4

$$\frac{\partial^2 E_p}{\partial \varphi^2} = 2\left( mgR - \frac{3mgR^2}{4l} \right)$$

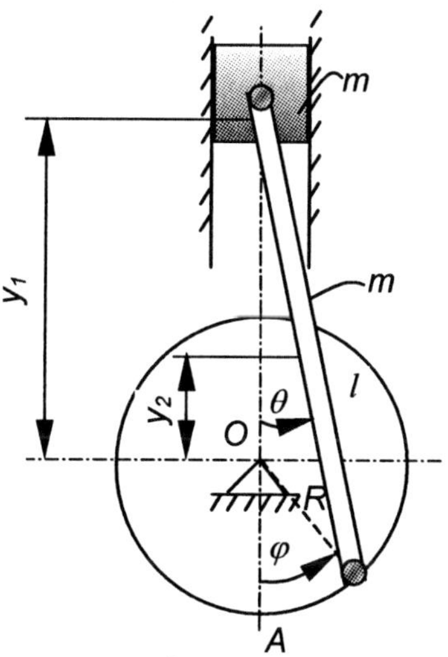


*Figure 3.5.*

$$2\left( mgR - \frac{3mgR^2}{4l} \right) > 0$$

$$2mgR - \frac{3}{2}\frac{mgR^2}{l} > 0$$

$$2 - \frac{3R}{2l} > 0$$

$$4l > 3R$$

$$\frac{4}{3}l > R$$

The last expression is the condition of stability.

3.3. Unextensible rope is linked to point *A* of lever *OA*. The lever has length *l* and weight *G*. The length of the rope is *2l*. The rope is thrown over the roller and Its dimensions are neglected. Weight *Q* is hooked up at the other end of the rope. The center of roller B and point *O* are on the same vertical, and *OB=l*. Find positions of equilibrium and check their stability. The frictions and mass of the rope are neglected.

Solution:
Total potential energy is:

$$E_p = E_{pl} + E_{pQ}$$

$E_{pl}$ is the potential energy of the lever, and $E_{pQ}$ equals potential energy of the weight

$$E_{pstapa} = Gy_1 = G\frac{l}{2}\cos\varphi$$

$$E_{pQ} = Qy_2 = Q\left(l - \overline{BC}\right)$$

Length of the rope (in Figure 3.6.):

$$l = \overline{AB} + \overline{BC}$$

$$\overline{BC} = l - \overline{AB} = l - 2l\sin\frac{\varphi}{2}$$

Potential energy:

$$E_p = \frac{Gl}{2}\cos\varphi + Q\left(l - l + 2l\sin\frac{\varphi}{2}\right) = \frac{Gl}{2}\cos\varphi + 2Ql\sin\frac{\varphi}{2}$$

$$\frac{\partial E_p}{\partial\varphi} = 0{:} -\frac{Gl}{2}\sin\varphi + Ql\cos\frac{\varphi}{2} = 0$$

We can use the formula:

$$\sin 2\frac{\varphi}{2} = 2\sin\frac{\varphi}{2}\cos\frac{\varphi}{2}$$

$$-\frac{Gl}{2}2\sin\frac{\varphi}{2}\cos\frac{\varphi}{2} + Ql\cos\frac{\varphi}{2} = 0$$

$$\cos\frac{\varphi}{2}\left(Ql - Gl\sin\frac{\varphi}{2}\right) = 0 \quad \Rightarrow \quad \cos\frac{\varphi_1}{2} = 0$$

$$\frac{\varphi_1}{2} = \arccos 0$$

$$\frac{\varphi_1}{2} = 90^\circ$$

First position of equilibrium: $\varphi_1$=180°.
Second position:

$$Ql - Gl\sin\frac{\varphi_2}{2} = 0$$

$$Ql = Gl\sin\frac{\varphi_2}{2}$$

$$\sin\frac{\varphi_2}{2} = \frac{Q}{G}$$

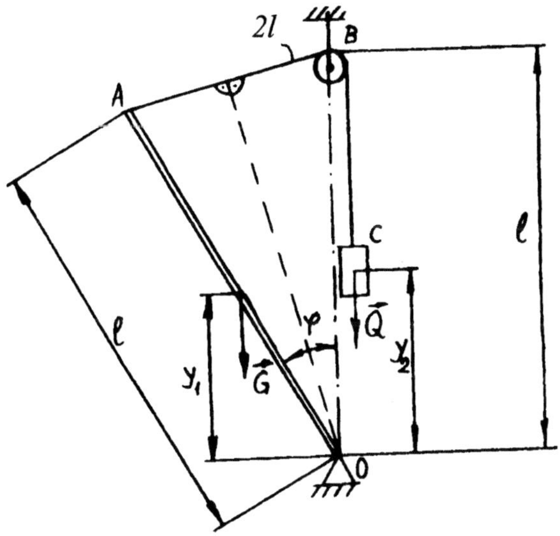


*Figure 3.6.*

The condition of stability:

$$\frac{\partial^2 E_p}{\partial\varphi^2} = -\frac{Gl}{2}\cos\varphi - \frac{1}{2}Ql\sin\frac{\varphi}{2}$$

For the first position ($\varphi$=$\varphi_1$):

$$\left(\frac{\partial^2 E_p}{\partial\varphi^2}\right)_{(\varphi=\varphi_1)} = -\frac{Gl}{2}\cdot(-1) - \frac{Ql}{2}\cdot 1$$

$$-\frac{Gl}{2}\cdot(-1) - \frac{Ql}{2}\cdot 1 > 0$$

$$\frac{Gl}{2} > \frac{Ql}{2}$$

$$G > Q$$

We use the next formula, for giving the second position:

$$\sin\frac{\varphi_2}{2} = \sqrt{\frac{1-\cos\varphi_2}{2}} \quad \Rightarrow \quad \cos\varphi_2 = 1 - 2\sin^2\frac{\varphi_2}{2} = 1 - 2\frac{Q^2}{G^2}$$

$$\left(\frac{\partial^2 E_p}{\partial\varphi^2}\right)_{(\varphi=\varphi_2)} = -\frac{Gl}{2}\left(1 - 2\frac{Q^2}{G^2}\right) - \frac{1}{2}Ql\frac{Q}{G}$$

$$-\frac{Gl}{2}\left(1 - 2\frac{Q^2}{G^2}\right) - \frac{1}{2}Ql\frac{Q}{G} > 0$$

After mathematical manipulation in the last expression, we get: $Q>G$.

If $Q>G$, then $\sin\frac{\varphi_2}{2} = \frac{Q}{G}$ is greater than 1, but the value of the function sine can't be greater than 1, so the position for $\varphi_2$ is not stable. The position of stable equilibrium is $\varphi_1$=180° with condition: $G>Q$.

3.4. Homogeneous square *AOB* is composed of two levers. The weight of each lever is *G*, and the length *AO*=*BO*=*l*. The angle ⊲*AOB* is 90°. The square rotates around the axis in point *O*. The axis is perpendicular with the plane of sketch (Figure 2.7.). There is weight *Q* in point *A* of lever *OA*. The spring with rigidity *c* holds the square in the position of equilibrium. Find the condition for equilibrium's position of stability (Figure 2.7.).

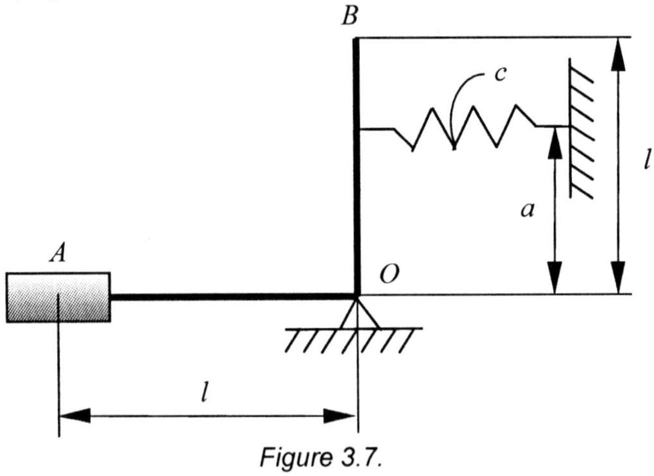


*Figure 3.7.*

Final solution:

$$a > \sqrt{\frac{Gl}{2c}}$$

3.5. There is a pendulum for the registration of horizontal vibrations in Figure 3.8. The pendulum is composed of levers *OA* and *BD*, cylindrical guide *C* and concentrated mass *B*. The lever *OA* rotates around the axis in point *O*. The cylindrical guide rotates around the axis in point *C*. The lever *BD* slides into guide *C*.

Find the condition in which the vertical position of the pendulum is stable. All frictions and the weight of the levers are neglected.

It is given:

$\overline{OA} = r$; $\overline{BA} = b$; $\overline{OC} = h$; weight of mass *B* is *G*.

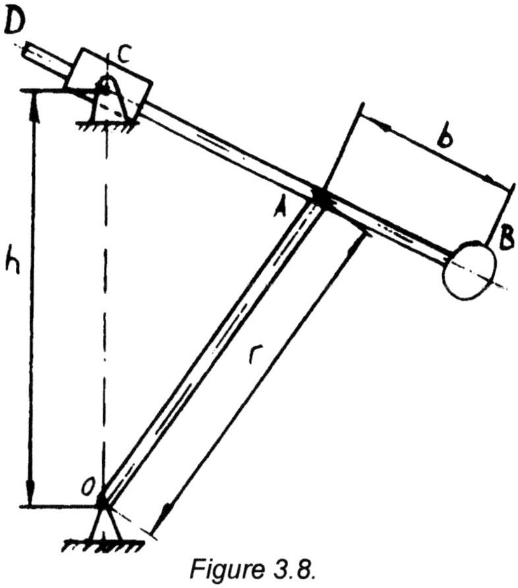


*Figure 3.8.*

Final solution:

$$h - r < \sqrt{rb}$$

# CHAPTER FOUR

## FREE DAMPED VIBRATIONS WITH ONE DEGREE OF VIBRATION'S FREEDOM

### 4.1. Short Theory Review

When resistances to motion aren't neglectable, we must take the force of resistance $F_w$ into consideration, and put it in the mathematical model. The force of resistance is proportionate to the speed of motion in the case of viscous friction:

$$F_W = k \cdot v \text{, namely } F_W = k \cdot \dot{q} \text{ , } \quad \ldots\ldots\ldots (4.1.)$$

where $k$ is the proportionality coefficient of $F_w$, $q$ is the generalized coordinate. The sense of force $F_w$ is always opposite motion's sense.

The general form of the differential equation for free damped vibrations with one degree of vibration's freedom:

$$\ddot{q} + 2n\dot{q} + \omega^2 q = 0 \text{ , } \quad \ldots\ldots\ldots\ldots\ldots (4.2.)$$

where $q$ is a generalized coordinate, $n$ is a factor of damping, and $w$ is a circular frequency. The values of n and $w$ depend upon the parameters of vibration's system.

The solution in equation (4.2.) has the form: $q = e^{rt}$, and the characteristic equation is:

$$r^2 + 2nr + \omega^2 = 0 \quad \ldots\ldots\ldots\ldots\ldots\ldots (4.3.)$$

It has solutions:

$$r_{1,2} = -n \pm \sqrt{n^2 - \omega^2} \quad \ldots\ldots\ldots\ldots (4.4.)$$

There are 3 cases. They depend upon $n$ and $\omega$:

#### 4.1.1. $n < \omega$

This is the case of low damping - the system does falling vibrations.

Solution of differential equation (4.2.):

$$q = Ae^{-nt} \sin\left(\sqrt{\omega^2 - n^2} \cdot t + \alpha\right) \text{, } \quad \ldots\ldots (4.5.)$$

where $A$ is amplitude of free damped vibrations. We can calculate it from initial conditions. The frequency of damped vibrations is $p$ and:

$$p=\sqrt{\omega^2-n^2}\ \text{[Hz]}$$

The initial lagging of this phase is $\alpha$ (it depends upon the initial conditions).

The period of damped vibrations:

$$T_W=\frac{2\pi}{p}[s] \qquad \ldots\ldots\ldots\ldots\ldots\ldots (4.6.)$$

Undimensional factor of damping:

$$\psi=\frac{n}{\omega} \qquad \ldots\ldots\ldots\ldots\ldots\ldots (4.7.)$$

The decrement of vibrations:

$$\Delta=e^{nTw} \qquad \ldots\ldots\ldots\ldots\ldots\ldots (4.8.)$$

The logarithmic decrement of vibrations:

$$D=\ln\Delta=nT_W \qquad \ldots\ldots\ldots\ldots\ldots\ldots (4.9.)$$

From (4.9.) it is possible to get the factor of damping $n$:

$$n=\frac{\ln\Delta}{Tw} \qquad \ldots\ldots\ldots\ldots\ldots\ldots (4.10.)$$

4.1.2. $n > \omega$

This is the case of high damping - the system does aperiodic motion.

The general solution of differential equation (4.2.):

$$x=Ae^{-nt}\cdot sh\left(\sqrt{n^2-\omega^2}\cdot t+\alpha\right) \qquad \ldots\ldots\ldots (4.11.)$$

4.1.3. $n = \omega$

This is the case of boundary aperiodic motion.

The general solution of differential equation (4.2.):

$$x=e^{-nt}\left(C_1+C_2t\right) \qquad \ldots\ldots\ldots\ldots\ldots\ldots (4.12.)$$

It is possible to use LaGrange's equation of second order for giving the differential equation (4.2.). LaGrange's equation for free damped vibrations with one degree of vibration's freedom has this form:

$$\frac{d}{dt}\left(\frac{\partial E_k}{\partial\dot{q}}\right)-\frac{\partial E_k}{\partial q}+\frac{\partial\Phi}{\partial\dot{q}}+\frac{\partial E_p}{\partial q}=0, \qquad \ldots\ldots (4.13.)$$

where $\phi$ is Rayleigh's function of system's energy

dissipation:

$$\Phi = \frac{1}{2} k \cdot \dot{q}^2 , \qquad \ldots\ldots\ldots\ldots\ldots\ldots \text{(4.14.)}$$

$k$ is coefficient of proportionality for resistance's force.

## 4.2. Problems

4.1. Vibration's system is composed of lever and mass $m$ (Figure 4.1.). Mass of the lever is neglected and it rotates without friction round the axis in point O. There is a muffler situated on the end of the lever, which has resistance's force $F_w$. The resistance's force is proportionate to speed and coefficient of the proportionality is $k$. Estimate the undimensional factor of damping, the decrement of vibrations and the logarithmic decrement of vibrations. Find the law of vibrations using the law of change of angular momentum. It is known: $a$=0,5[m]; $b$=0,3[m]; $m$=10[kg]; $k$=200[Ns/m] and the initial conditions:

$$\varphi_0 = 0; \dot{\varphi}_0 = 1\left[\frac{rad}{s}\right]; t_0 = 0[s]$$

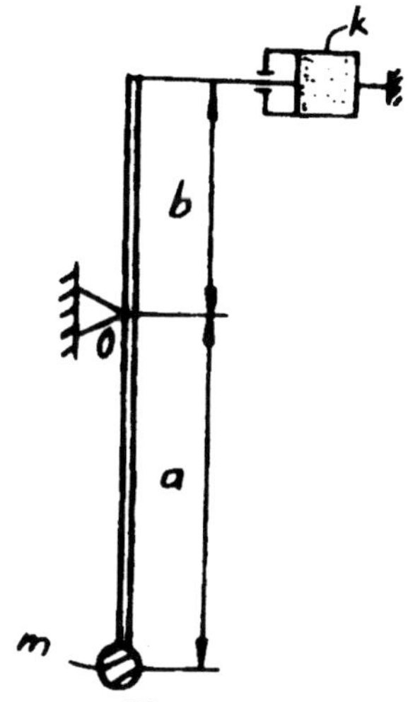


*Figure 4.1.*

Solution:

The law of change of angular momentum (Figure 4.2.):

$$J_0\ddot{\varphi} = \sum M_O^{Fi}$$

$$ma^2\ddot{\varphi} = -F_w b\cos\varphi - mga\sin\varphi$$

$$ma^2\ddot{\varphi} + k\dot{\varphi}b^2 \cdot 1 + mga\varphi = 0$$

$$\ddot{\varphi} + \frac{kb^2}{ma^2}\dot{\varphi} + \frac{g}{a}\varphi = 0 \qquad \ldots\ldots\ldots\ldots\ldots (4.2.1.)$$

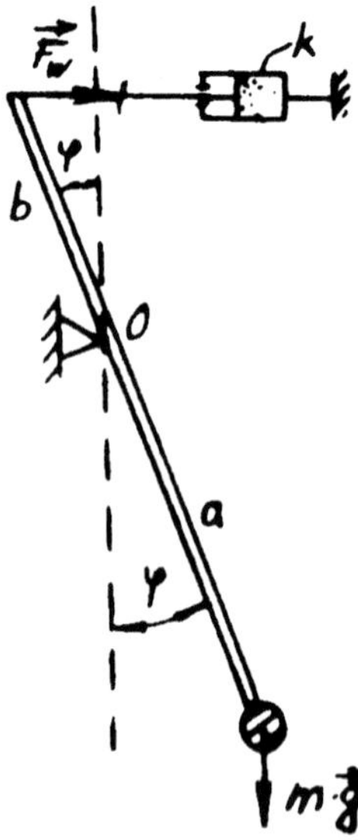

*Figure 4.2.*

We compare the differential equation (4.2.1.) with a general form (4.2.):

$$2n = \frac{kb^2}{ma^2} \quad \Rightarrow \quad n = \frac{kb^2}{2ma^2}$$

$$\omega^2 = \frac{g}{a} \quad \Rightarrow \quad \omega = \sqrt{\frac{g}{a}}$$

$n$=3,6[s$^{-1}$];

$\omega$=4,43[s$^{-1}$].

We can see that $n<\omega$. It is the case of low damping.

$$p = \sqrt{\omega^2 - n^2} = \sqrt{\left(4{,}43\left[s^{-1}\right]\right)^2 - \left(3{,}6\left[s^{-1}\right]\right)^2} = 2{,}58\left[s^{-1}\right]$$

The undimensional factor of damping (4.7.):

$$\psi = \frac{n}{\omega} = \frac{3,6}{4,43} = 0,813$$

The decrement of vibrations (4.8.):

$$\Delta = e^{nTw} = e^{3,6 \cdot 2,435} = 6412,47,$$

where $T_W = \frac{2\pi}{p} = \frac{2\pi}{2,58} = 2,435[s]$

The logarithmic decrement of vibrations (4.9.):

$D$=ln$\Delta$=$nT_W$= $3,6 \cdot 2,435 = 8,766$

The solution has the form:

$$\varphi = Ae^{-nt}\sin(pt + \alpha)$$

$$\dot{\varphi} = Ae^{-nt}(-n)\sin(pt + \alpha) + Ae^{-nt}p\cos(pt + \alpha)$$

$$\varphi_0 = 0;\, t_0 = 0[s] \quad \Rightarrow \quad 0 = Ae^{-3,6 \cdot 0}\sin(2,58 \cdot 0 + \alpha),$$

$$0 = A\sin\alpha \quad \Rightarrow \quad \sin\alpha = 0 \quad \Rightarrow \quad \alpha = 0$$

$$\dot{\varphi}_0 = 1\left[\frac{m}{s}\right];\, t_0 = 0[s] \quad \Rightarrow \quad 1 = Ae^{-3,6 \cdot 0} \cdot (-3,6) \cdot \sin(2,58 \cdot 0 + 0) +$$

$$+ Ae^{-3,6 \cdot 0} \cdot 2,58 \cdot \cos(2,58 \cdot 0 + 0),$$

$$1 = A \cdot 1 \cdot 2,58 \cdot 1 \quad \Rightarrow \quad A = 0,39[m]$$

The law of vibrations:

$$\varphi = 0,39e^{-3,6 \cdot t} \cdot \sin 2,58\ \text{t} \quad \text{[m,s]}$$

4.2. Estimate the differential equation for vibration's system in the previous problem (Figure 4.1.) using d'Alembert's law.

Solution:

We have to add the inertial forces. Each mass in the system has the inertial force. Here, we have just one mass and inertial force (Figure 4.3.)

$$\sum M_0 = 0$$

$$F_W b\cos\varphi + mga\sin\varphi + F_T^{in}a = 0 \qquad \text{...... (4.2.2.)}$$

Tangential component:

$$F_T^{in} = m\frac{d^2s}{dt} = m\frac{dv}{dt} = m\frac{d}{dt}(a\dot{\varphi}) = ma\ddot{\varphi}$$

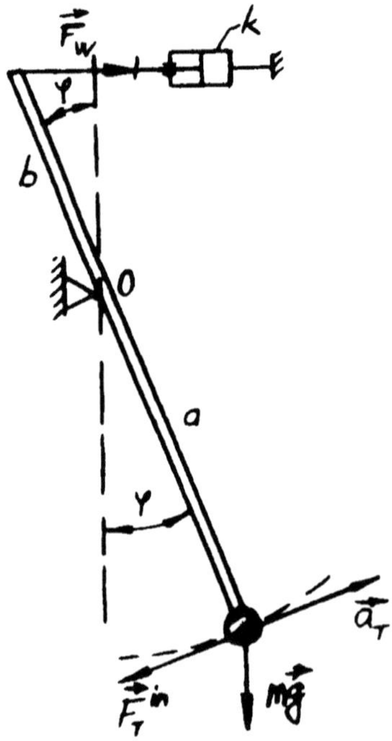


*Figure 4.3.*

Put it in the equation (4.2.2.):

$$kb^2\dot{\varphi}\cos\varphi + mga\sin\varphi + ma^2\ddot{\varphi} = 0$$

For small vibrations: $\sin\varphi \approx \varphi$ and $\cos\varphi \approx 1$

$$ma^2\ddot{\varphi} + kb^2\dot{\varphi} + mga\varphi = 0$$

$$\ddot{\varphi} + \frac{kb^2}{ma^2}\dot{\varphi} + \frac{g}{a}\varphi = 0$$

We get the same differential equation as (4.2.1.) in the previous problem.

4.3. Find the differential equation for vibration's system in the problem 4.1. (Figure 4.1.) by using LaGrange's equation.

Solution:

$$E_k = \frac{1}{2}mv_m^2 = \frac{1}{2}ma^2\dot{\varphi}^2; \quad \Phi = \frac{1}{2}kb^2\dot{\varphi}^2; \quad E_p = mg(a - a\cos\varphi)$$

When we use LaGrange's equation it is necessary to approximate $\cos\varphi$ with $(1-\varphi^2/2)$ and $\sin\varphi$ with $\varphi$ for small vibrations:

$$E_p \approx mg\left(a - a + a\frac{\varphi^2}{2}\right) = mga\frac{\varphi^2}{2}$$

LaGrange's equation (4.13.):

$$\frac{d}{dt}\left(\frac{\partial E_k}{\partial \dot{\varphi}}\right) - \frac{\partial E}{\partial \varphi} + \frac{\partial E_p}{\partial \varphi} + \frac{\partial \Phi}{\partial \dot{\varphi}} = 0$$

$$ma^2\ddot{\varphi} - 0 + kb^2\dot{\varphi} + mga\varphi = 0 \quad \Rightarrow \quad \ddot{\varphi} + \frac{kb^2}{a^2 m}\dot{\varphi} + \frac{g}{a}\varphi = 0$$

The last equation is the same as (4.2.1.).

4.4. There is a homogeneous disc with mass *m* and radius *r*, inside a cylindrical half-roll. The radius of the half-roll is *R*. There is a resistance of the disc's vibrations, and it is taken into consideration with a muffler (Figure 4.4.). The muffler has coefficient *k*. Calculate the factor of damping *n* and the circular frequency $\omega$.

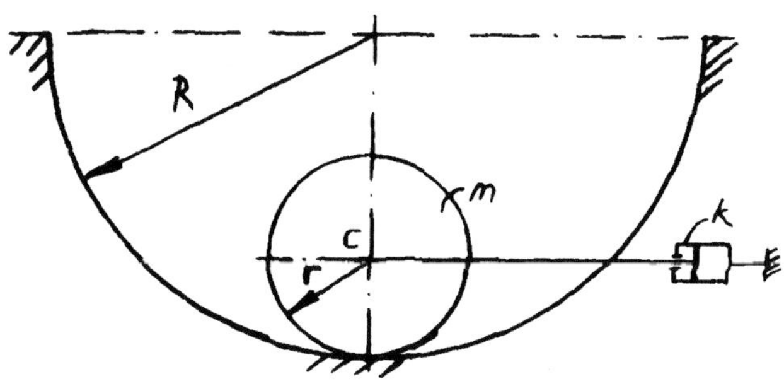


*Figure 4.4.*

Solution:

The disc does plane movement and the kinetic energy ($E_k$) is composed of kinetic energy of translation ($E_{ktran}$) and rotation ($E_{krot}$).

$$E_k = E_{krot} + E_{ktran} = \frac{1}{2}mv_c^2 + \frac{1}{2}J_c\dot{\theta}^2$$

$$v_c = (R - r)\dot{\varphi}^2; \quad J_c = \frac{1}{2}mr^2$$

$$E_k = \frac{1}{2}m(R-r)^2\dot{\varphi}^2 + \frac{1}{2}\frac{1}{2}mr^2\left(\frac{R}{r}-1\right)^2\dot{\varphi}^2$$

$\dot{\theta} = \dfrac{d\theta}{dt}$ is got in Figure 4.5:

$$\widehat{KP} = \widehat{SP} \quad \Leftrightarrow \quad r(\varphi+\theta) = R\varphi$$

$$\varphi + \theta = \frac{R}{r}\varphi$$

$$\theta = \left(\frac{R}{r}-1\right)\varphi \Bigg/ \frac{d}{dt} \quad \Leftrightarrow \quad \dot{\theta} = \left(\frac{R}{r}-1\right)\dot{\varphi}$$

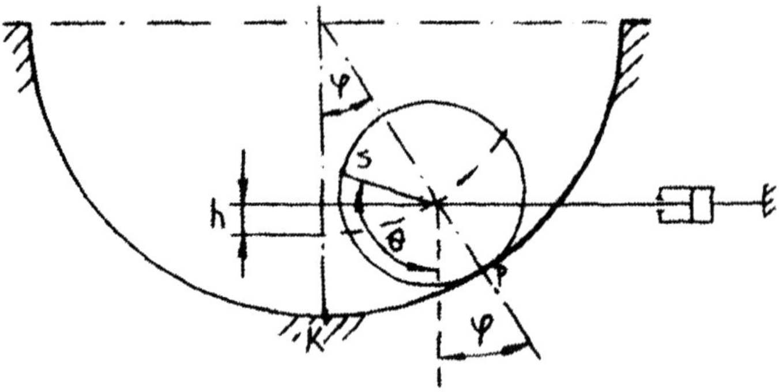


*Figure 4.5.*

The potential energy:

$$E_p = mgh = mg\cdot\left[(R-r)-(R-r)\cos\varphi\right] \approx$$

$$\approx mg\left[(R-r)\left(1-1+\frac{\varphi^2}{2}\right)\right] =$$

$$= mg\left(R\frac{\varphi^2}{2} - r\frac{\varphi^2}{2}\right)$$

The function of system's energy dissipation:

$$\Phi = \frac{1}{2}kv_c^2 = \frac{1}{2}k(R-r)^2\dot{\varphi}^2$$

LaGrange's equation:

$$\frac{d}{dt}\left(\frac{\partial E_k}{\partial \dot{\varphi}}\right)-\frac{\partial E}{\partial \varphi}+\frac{\partial E_p}{\partial \varphi}+\frac{\partial \Phi}{\partial \dot{\varphi}}=0$$

$$m(R-r)^2\ddot{\varphi}+\frac{1}{2}mr^2\left(\frac{R}{r-1}\right)^2\ddot{\varphi}+k(R-r)^2\dot{\varphi}+(mgR-mgr)\varphi=0$$

$$\ddot{\varphi}+\frac{2k}{3m}\dot{\varphi}+\frac{2g}{3(R-r)}\varphi=0$$

Compare the last differential equation with (4.2.):

$$2n=\frac{2k}{3m}\quad\Rightarrow\quad n=\frac{k}{3m},\qquad \omega^2=\frac{2g}{3(R-r)}\quad\Rightarrow\quad \omega=\sqrt{\frac{2g}{3(R-r)}}$$

4.5. It is necessary to write the differential equation of small vibrations, the circular frequency $\omega$, the factor of damping $n$, the frequency of damped vibration $p$ and the period of damped vibration $T_W$ for vibration's system in Figure 4.6.a). It is given: $k$=250kg/s; $a$=0,5m; $m$=1,75kg; $l$=0,75m; $c$=2120N/m.

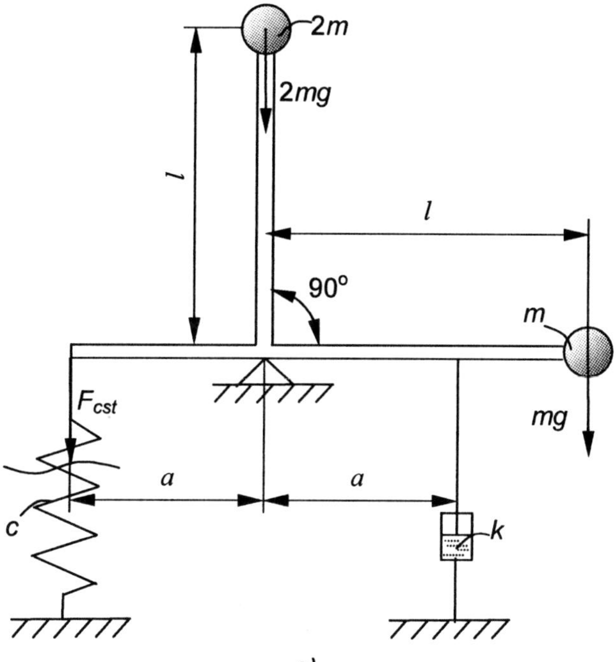


a)
*Figure 4.6.*

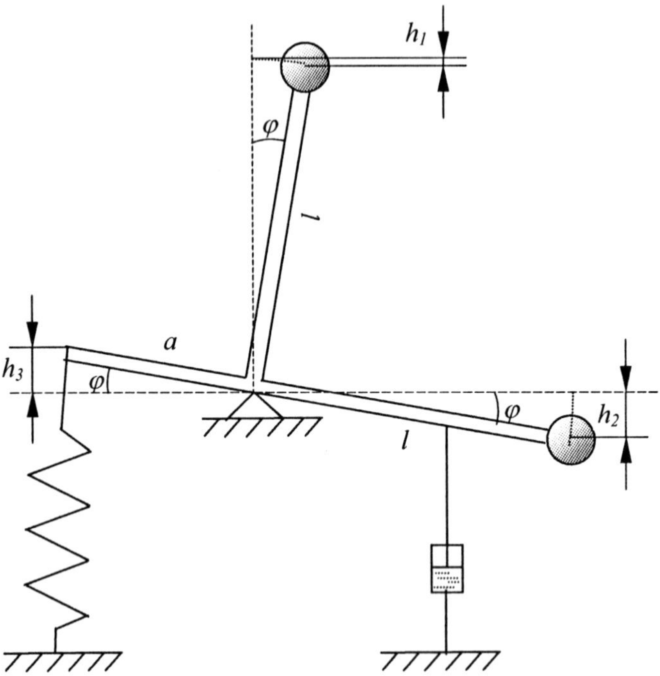


b)

*Figure 4.6.*

Solution:

$$E_k = E_{km} + E_{k2m} = \frac{1}{2}ml^2\dot{\varphi}^2 + \frac{1}{2}2ml^2\dot{\varphi}^2$$

$$E_k = \frac{3}{2}ml^2\dot{\varphi}^2$$

The potential energy of system (Figure 4.6.b)):

$$E_p = E_{pm} + E_{p2m} + E_{pc} = -mgh_2 - 2mgh_1 + \frac{1}{2}c(f_{st} + h_3)^2$$

$$h_1 = l - l\cos\varphi\,;\ h_2 = l\sin\varphi\,;\ h_3 = a\sin\varphi$$

$$E_p \approx -mgl\varphi - 2mgl\left(1 - 1 + \frac{\varphi^2}{2}\right) + \frac{1}{2}c(f_{st} + a\varphi)^2$$

$$\Phi = \frac{1}{2}ka^2\dot{\varphi}^2$$

LaGrange's equation (4.13.):

$$\frac{d}{dt}\left(\frac{\partial E_k}{\partial \dot{q}}\right) - \frac{\partial E_k}{\partial q} + \frac{\partial \Phi}{\partial \dot{q}} + \frac{\partial E_p}{\partial q} = 0$$

$$3ml^2\ddot{\varphi} + ka^2\dot{\varphi} - mgl - 2mgl\varphi + c(f_{st} + a\varphi)a = 0$$

$$3ml^2\ddot{\varphi} + ka^2\dot{\varphi} - 2mgl\varphi + ca^2\varphi = mgl - caf_{st}$$

The static condition of equilibrium in Figure 4.6.a):

$$\sum M_o = 0$$

$$F_{st}a - mgl = 0$$

$$caf_{st} - mgl = 0$$

Put in the differential equation:

$$3ml^2\ddot{\varphi} + ka^2\dot{\varphi} - 2mgl\varphi + ca^2\varphi = 0$$

$$\ddot{\varphi} + \frac{ka^2}{3ml^2}\dot{\varphi} + \frac{ca^2 - 2mgl}{3ml^2}\varphi = 0$$

$$2n = \frac{ka^2}{3ml^2} \doteq \frac{250\left[\frac{kg}{s}\right]\cdot 0{,}5^2\left[m^2\right]}{3\cdot 1{,}75[kg]\cdot 0{,}75^2\left[m^2\right]} = 21{,}16\left[\frac{rad}{s}\right]$$

$$n = 10{,}58\left[\frac{rad}{s}\right]$$

$$\omega^2 = \frac{ca^2 - 2mgl}{3ml^2}$$

$$\omega = \sqrt{\frac{ca^2 - 2mgl}{3ml^2}} \quad \Rightarrow \quad \omega = 13{,}07\left[\frac{rad}{s}\right]$$

$$p = \sqrt{\omega^2 - n^2} = 7{,}67\left[\frac{rad}{s}\right]$$

$$T_W = \frac{2\pi}{p} = \frac{2\pi}{7{,}67\left[\frac{rad}{s}\right]} = 0{,}81[s]$$

4.6. A rigid lever has two concentrated masses 2*m* and *m* on distance *l* from the point *O* (Figure 4.7.). The lever rotates around this axis in point *O* without friction. There is a spring with rigidity *c* and a muffler with coefficient *k* from *a* to point *O*. Calculate the circular frequency $\omega$ and the factor of damping *n* using D'Alembert's law.

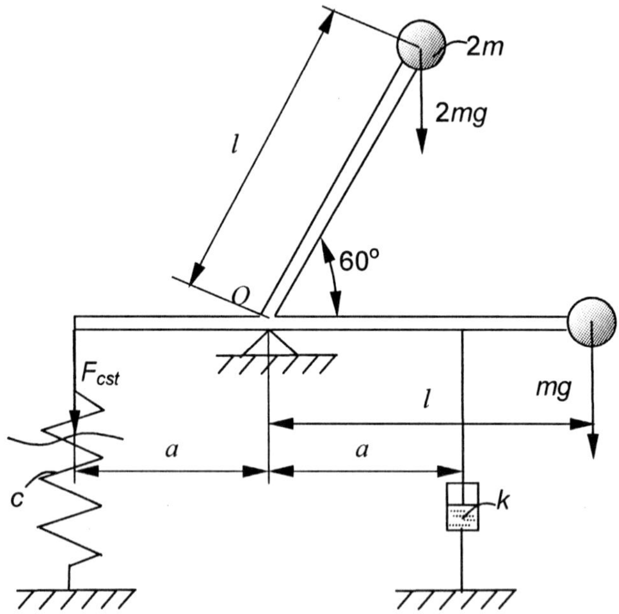


Figure 4.7.

Solution:

The static condition of equilibrium (Figure 4.7.):

$$\sum M_o = 0$$

$$2mgl \cdot \cos 60^o + mgl - F_{st} a = 0$$

$$2mgl \cdot \frac{1}{2} + mgl - cf_{st} a = 0$$

$$2mgl - caf_{st} = 0 \quad \ldots\ldots\ldots\ldots\ldots\ldots\ldots \quad (4.2.3.)$$

It is necessary to add the inertial forces and displace the masses off the position of equilibrium (Figure 4.8.):

$$\sum M_o = 0$$

$$2mgl \cdot \cos(60^o - \varphi) + mgl \cos\varphi - F_c a \cos\varphi -$$

$$- F_W a \cos\varphi - (F_T^{in})_m l - (F_T^{in})_{2m} l = 0$$

The intensity of tangential inertial forces:

$$(F_T^{in})_m = ma_T = m\frac{dv}{dt} = ml\ddot{\varphi}$$

$$(F_T^{in})_{2m} = 2ml\ddot{\varphi}$$

The force of spring:

$$F_c = c\cdot\left(f_{st} + a\sin\varphi\right) = cf_{st} + ca\sin\varphi$$

Put it in the differential equation:

$$2mgl\cdot\cos(60^o - \varphi) + mgl\cos\varphi - cf_{st}a\cos\varphi -$$

$$- a^2c\cos\varphi\sin\varphi - kv_Aa\cos\varphi - ml^2\ddot{\varphi} - 2ml^2\ddot{\varphi} = 0$$

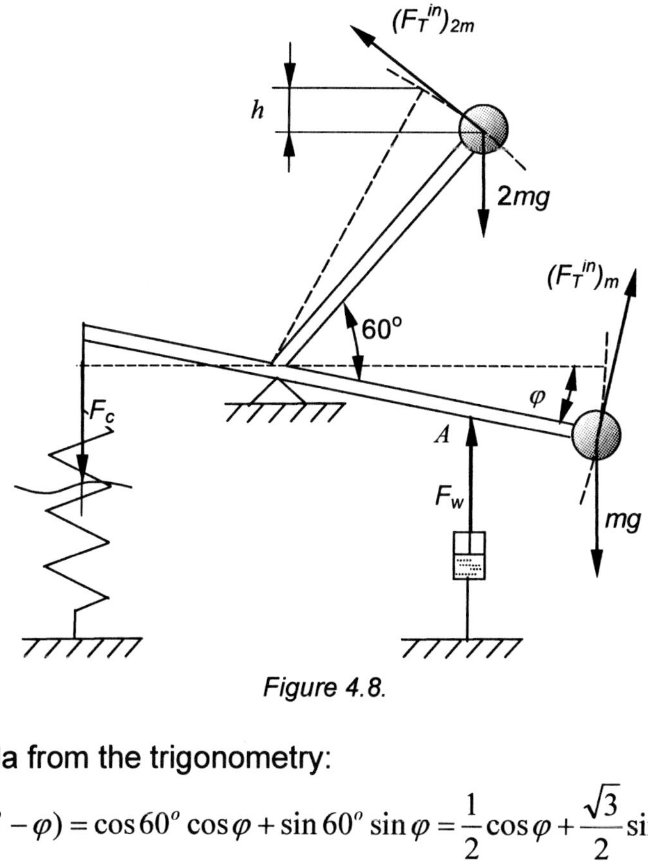


Figure 4.8.

Formula from the trigonometry:

$$\cos(60^o - \varphi) = \cos 60^o \cos\varphi + \sin 60^o \sin\varphi = \frac{1}{2}\cos\varphi + \frac{\sqrt{3}}{2}\sin\varphi$$

For the small vibration $\sin\varphi\approx\varphi$ and $\cos\varphi\approx 1$:

$$2mgl + \sqrt{3}mgl\varphi - cf_{st}a - a^2c\varphi - ka^2\dot{\varphi} - 3ml^2\ddot{\varphi} = 0$$

Use the static condition of equilibrium (4.2.3.):

$$\sqrt{3}mgl\varphi - ca^2\varphi - ka^2\dot{\varphi} - 3ml^2\ddot{\varphi} = 0 \qquad \text{. . . . (4.2.4.)}$$

$$\ddot{\varphi} + \frac{ka^2}{3ml^2}\dot{\varphi} + \frac{ca^2 - \sqrt{3}mgl}{3ml^2}\varphi = 0$$

$$2n = \frac{ka^2}{3ml^2} \quad \Rightarrow \quad n = \frac{ka^2}{6ml^2}$$

$$\omega^2 = \frac{ca^2 - \sqrt{3}mgl}{3ml^2} \quad \Rightarrow \quad \omega = \sqrt{\frac{ca^2 - \sqrt{3}mgl}{3ml^2}}$$

4.7. Estimate the differential equation for vibration's system in the previous problem (Figure 4.7.) using LaGrange's equation.

Solution:
In Figure 4.8:

$$h = l\sin 60^o - l\sin(60^o - \varphi)$$

$$h = l\frac{\sqrt{3}}{2} - l(\sin 60^o \cos\varphi - \cos 60^o \sin\varphi)$$

When we use LaGrange's equation, it is necessary to approximate $\cos\varphi \approx 1 - \frac{\varphi^2}{2}$ and $\sin\varphi \approx \varphi$ for small vibrations:

$$h = l\frac{\sqrt{3}}{2} - l\left(\frac{\sqrt{3}}{2} - \frac{\sqrt{3}}{2}\frac{\varphi^2}{2} - \frac{\varphi}{2}\right)$$

$$h = \frac{\sqrt{3}}{4}l\varphi^2 + \frac{l}{2}\varphi$$

The kinetic ($E_k$) and potential ($E_p$) energy:

$$E_k = \frac{1}{2}mv_m^2 + \frac{1}{2}2mv_{2m}^2 = \frac{1}{2}ml^2\dot{\varphi}^2 + ml^2\dot{\varphi}^2 = \frac{3}{2}ml^2\dot{\varphi}^2$$

$$E_p = -mgl\sin\varphi - 2mgh + \frac{1}{2}c(f_{st} + a\cdot\sin\varphi)^2 \approx$$

$$\approx -mgl\varphi - mg\frac{\sqrt{3}}{2}l\varphi^2 - mgl\varphi + \frac{1}{2}c(f_{st} + a\cdot\varphi)^2$$

The function of the system's energy dissipation:

$$\Phi = \frac{1}{2}kv_A^2 = \frac{1}{2}k(a\dot{\varphi})^2 = \frac{1}{2}ka^2\dot{\varphi}^2$$

LaGrange's equation (4.13.):

$$\frac{d}{dt}\left(\frac{\partial E_k}{\partial \dot{q}}\right)-\frac{\partial E_k}{\partial q}+\frac{\partial \Phi}{\partial \dot{q}}+\frac{\partial E_p}{\partial q}=0$$

$$3ml^2\ddot{\varphi}+ka^2\dot{\varphi}-2mgl-\sqrt{3}mgl\varphi+caf_{st}+ca^2\varphi=0$$

Put the static condition of equilibrium (4.2.3.) from the previous problem in the last differential equation:

$\sqrt{3}mgl\varphi-ca^2\varphi-ka^2\dot{\varphi}-3ml^2\ddot{\varphi}=0$ - it is equivalent equation to (4.2.4.)

4.8. Estimate vibration's law of systems in Figure 4.9. (Position of equilibrium is shown in the figure). The lever *AB* is swung around joint *O* for angle 6° in initial moment. The initial velocity is 0. All springs are unloaded in the position of equilibrium. The parameters of the system are:

$\frac{c}{m}=75$; $\frac{g}{l}=20$; $\frac{k}{m}=32$; $M_1=2\cdot M_2=6\cdot m$.

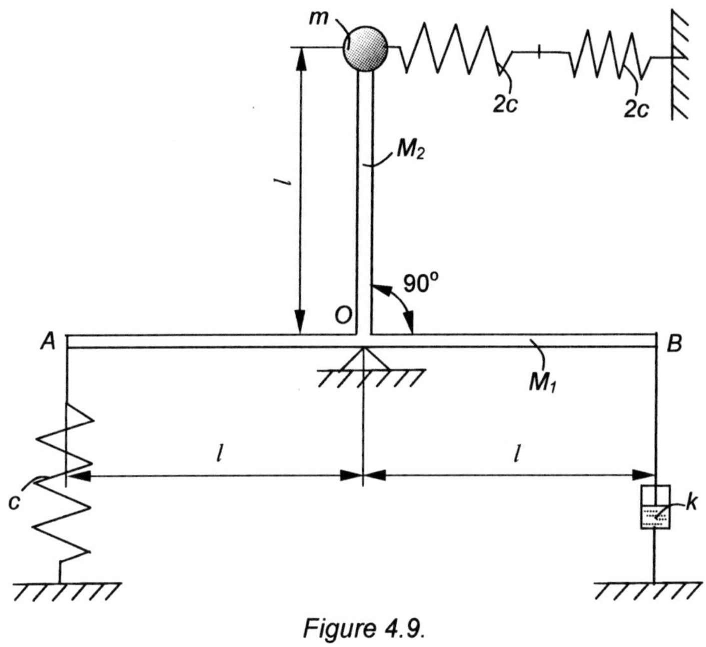


*Figure 4.9.*

Final solution:

$$\varphi=e^{-nt}\left(\frac{\pi}{30}\cos 3t+\frac{2\pi}{45}\sin 3t\right)$$

4.9. Calculate $\omega$, $T$ and the factor of damping $n$ for vibration's system shown in the position of equilibrium in Figure 4.10.
It is known:
$M$=9[kg]; $R$=0,4[m]; $c$=1000[N/m]; $a$=0,35[m]; $k$=200[Ns/m].

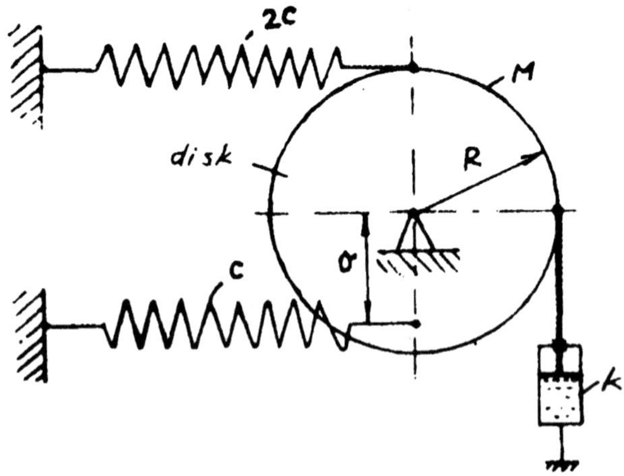


*Figure 4.10.*

Final solution:
$\omega$=22,79[rad/s];
$T$=0,253[s];
$n$=22,2[s$^{-1}$]

# CHAPTER FIVE

## FORCED UNDAMPED VIBRATIONS WITH ONE DEGREE OF VIBRATION'S FREEDOM

### 5.1. Short Theory Review

$F_\Omega$ is excitative force and it depends upon time ($F_\Omega = F_\Omega(t)$ ). If the force $F_\Omega$ acts upon the vibration's system, the vibrations are forced.
If the force $F_\Omega$ changes harmoniously, it has form:

$$F_\Omega(t) = F_0 \cos(\Omega t - \beta) \qquad \text{............(5.1.)}$$

where $F_o$ is the amplitude, $\Omega$ is the frequency, $(\Omega t-\beta)$ is the phase and $\beta$ is initial phase of excitative force.
General form of differential equation for forced undamped vibrations with one degree of vibration's freedom:

$$\ddot{q} + \omega^2 q = h\cos(\Omega t - \beta) \qquad \text{.......... (5.2.)}$$

where $h$ is the amplitude of excitation and $q$ is generalized coordinate.
It is a linear unhomogeneous differential equation of second order, and its general solution has the form:

$$q = q_h + q_p ,$$

where $q_h$ is a homogeneous and $q_p$ is particular part of the solution. The homogeneous part:
$q_h = C_1 \cos\omega t + C_2 \sin\omega t$ - It is solution (2.3.) of equation (2.1.).
The particular part must have the same form as the excitative force (if $F_\Omega$ has $\cos(\Omega t - \beta)$, $q_p$ must have $\cos(\Omega t - \beta)$), so:

$$q_p = \frac{h}{\omega^2 - \Omega^2}\cos(\Omega t - \beta)$$

$$q = C_1 \cos\omega t + C_2 \sin\omega t + \frac{h}{\omega^2 - \Omega^2}\cos(\Omega t - \beta) \quad \text{...(5.3.)}$$

Dynamic factor of vibration's increasing or Magnification factor (multiplier of resonance):

$$\eta = \frac{A}{A_{st}} = \frac{1}{1-\left(\frac{\Omega}{\omega}\right)^2} = \frac{1}{1-\psi^2} \qquad \text{. . . . . . . . (5.4.)}$$

It is the ratio of dynamic ($A$) and static ($A_{st}$) amplitude. If $\Omega=\omega$ then the resonance appears.

For forced undamped vibrations with one degree of vibration's freedom, LaGrange's equation has the form:

$$\frac{d}{dt}\left(\frac{\partial E_k}{\partial \dot{q}}\right) - \frac{\partial E_k}{\partial q} + \frac{\partial E_p}{\partial q} = Q_\Omega \qquad \text{. . . . . . . (5.5.)}$$

where $E_k$ is kinetic energy, $E_p$ is potential energy, $Q_\Omega$ is generalized excitative force (moment of force); $q$ is generalized coordinate (linear coordinate or angle).

## 5.2. Problems

5.1. Gunpowder's gas acts upon the piston of the shutter with force $F_\Omega$ on the occasion of firing a bullet's series. We approximate the force with $F_\Omega$=25sin(60$t$)[N,s], because of giving the simpler mathematical model. The spring of the shutter has rigidity $c$=2000[N/m]. Estimate the physical characteristics (circular frequency $\omega$, period $T$, frequency $\Omega$ and period $T_\Omega$ of the excitative force), dynamic factor of vibration's increasing $\eta$ and the law of shutter's vibrations, if the shutter is considered like concentrated mass (Figure 5.2.). All frictions are neglected.

Initial conditions: $t_0$=0[s]; $x_0$=0[m]; $\dot{x}_0 = 0$ [m/s].

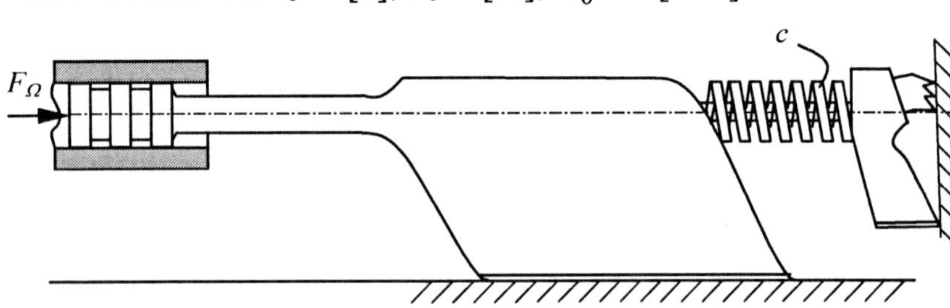


*Figure 5.1. Physical model*

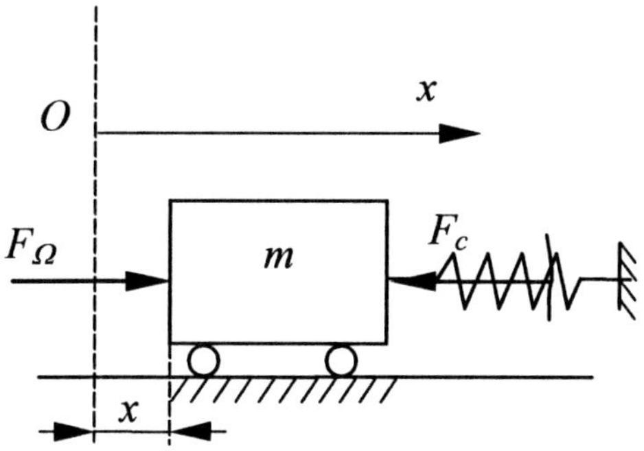


*Figure 5.2. Mechanical model*

Solution:
Newton's second law (Figure 5.2.):

$$m\ddot{x} = \sum F_{ix}$$

$$m\ddot{x} = F_\Omega - F_c$$

$$m\ddot{x} = F_\Omega - cx$$

$$m\ddot{x} + cx = F_\Omega$$

$$\ddot{x} + \frac{c}{m}x = \frac{25}{m}\sin(60t)$$

$$\Rightarrow \quad \omega = \sqrt{\frac{c}{m}} = \sqrt{\frac{2000\left[\frac{N}{m}\right]}{0{,}4[kg]}} = 70{,}71\left[s^{-1}\right]$$

$$T = \frac{2\pi}{\omega} = \frac{2\pi}{70{,}71\left[s^{-1}\right]} = 0{,}09[s];\ \Omega = 60\left[s^{-1}\right] \ \Rightarrow \ T_\Omega = \frac{2\pi}{\Omega} = 0{,}10[s]$$

The law of motion has form (5.3.):

$$x = A\cos\omega t + B\sin\omega t + \frac{h}{\omega^2 - \Omega^2}\cos\Omega t \ . \quad (5.2.1.)$$

$$h = \frac{F_0}{m} = \frac{25[N]}{0{,}4[kg]} = 62{,}5\left[\frac{m}{s}\right]$$

The constants *A* and *B* are calculated from initial conditions:

$$t_0 = 0[s];\ x_0 = 0[m];\ \dot{x}_0 = 0\left[\frac{m}{s}\right]$$

$$\dot{x} = -A\omega \sin \omega t + B\omega \cos \omega t - \frac{h\Omega}{\omega^2 - \Omega^2} \sin \Omega t \quad .......(5.2.2.)$$

Put the initial conditions in equations (5.2.1.) i (5.2.2.):

$$0 = A \cdot 1 + 0 + \frac{h}{\omega^2 - \Omega^2} \cdot 1 \Rightarrow A = -\frac{h}{\omega^2 - \Omega^2}$$

$$0 = -0 + \omega B \cdot 1 - \frac{h \cdot \Omega}{\omega^2 - \Omega^2} \cdot 0 \Rightarrow B = 0$$

The law of motion:

$$x = -\frac{h}{\omega^2 - \Omega^2} \cos \omega t + \frac{h}{\omega^2 - \Omega^2} \cos \Omega t =$$

$$= x = \frac{h}{\omega^2 - \Omega^2} (\cos \Omega t - \cos \omega t)$$

We can use the trigonometrical formula:

$$\cos \Omega t - \cos \omega t = -2 \sin\left(\frac{\Omega + \omega}{2} t\right) \sin\left(\frac{\Omega - \omega}{2} t\right)$$

Now the law has form:

$$x = -\frac{h}{\omega^2 - \Omega^2} 2 \sin\left(\frac{\Omega + \omega}{2} t\right) \sin\left(\frac{\Omega - \omega}{2} t\right)$$

$$x = -0{,}045 \cdot 2 \cdot \sin(65{,}36t) \sin(-5{,}36t)$$

From trigonometry sin ($-\alpha$)= - sin($\alpha$):

sin (-5,36 $t$)= - sin(5,36 $t$)

Finally the law: $x = 0{,}09 \sin(65{,}36 \cdot t) \cdot \sin(5{,}36 \cdot t)$

Dynamic factor of vibration's increasing (Magnification factor):

$$\eta = \frac{1}{1 - \left(\frac{\Omega}{\omega}\right)^2} = \frac{1}{1 - \left(\frac{60[s^{-1}]}{70{,}71[s^{-1}]}\right)^2} = 3{,}57;$$

$$\psi = \frac{\Omega}{\omega} = \frac{60[s^{-1}]}{70{,}71[s^{-1}]} = 0{,}85$$

There is the diagram of resonance in Figure 5.3. It is the graph of function $\eta = \frac{1}{1 - \psi^2}$. We get the value $\eta = 3{,}57$ for $\psi = 0{,}85$ in the diagram. For $\psi$=1 the value of $\eta$ is infinite and

it is the resonance. Then the amplitude is high and it is possible that an vibration's system will be broken.
Therefore, it is necessary to choose the parameters of the system (*c*, *m*) so that $\psi$ is less (range *A* in the diagram) or higher (range *B*). $\psi$ **is less** if $\Omega$ is less or $\omega$ is higher. $\psi$ **is higher** if $\Omega$ is higher or $\omega$ is less.

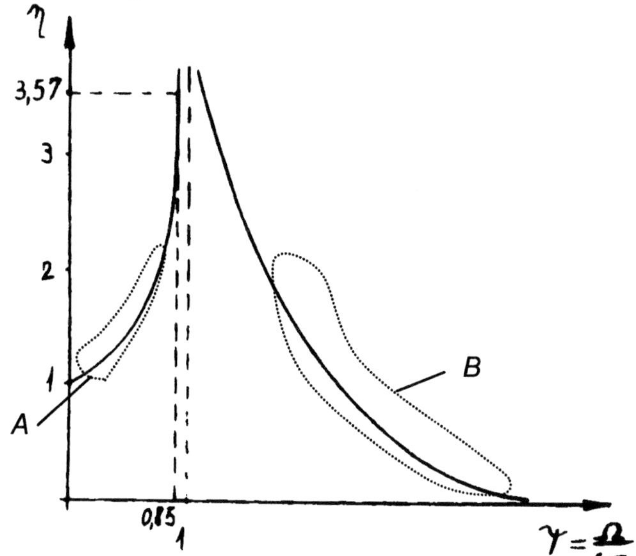


*Figure 5.3.*

5.2. Concentrated mass *m* is situated at the end of lever. Mass of the lever is neglected. Excitative force $F_\Omega = F_0 \sin \Omega t$ acts upon the lever. The spring is situated on distance $a$ from the link *O* (Figure 5.4.). Rigidity of the spring is *c*. Lever rotates round link *O* without friction. Write a differential equation of the system's vibration using LaGrange's equation of second order.

Solution:
Static condition of equilibrium (Figure 5.4.):

$$\sum M_o = 0$$

$$F_{cst} a - mgl = 0$$

$$c f_{st} a - mgl = 0 \quad \ldots\ldots\ldots\ldots\ldots\ldots \text{(5.2.3.)}$$

Kinetic and potential energy of the system:

$$E_k = \frac{1}{2} m v_m^2 = \frac{1}{2} m (l\dot{\varphi})^2$$

$$E_p = -mgl\varphi + \frac{1}{2} c (f_{st} + a\varphi)^2$$

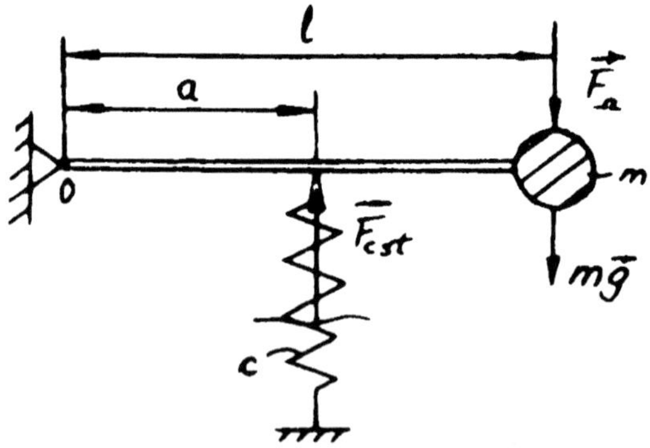


*Figure 5.4.*

LaGrange's equation:

$$\frac{d}{dt}\left(\frac{\partial E_k}{\partial \dot{\varphi}}\right) - \frac{\partial E_k}{\partial \varphi} + \frac{\partial E_p}{\partial \varphi} = M_\Omega$$

where $M_\Omega$ is excitative moment: $M_\Omega = F_\Omega l$

$$ml^2 \ddot{\varphi} + c(f_{st} + a\varphi)a - mgl = (F_o \sin \Omega t)l$$

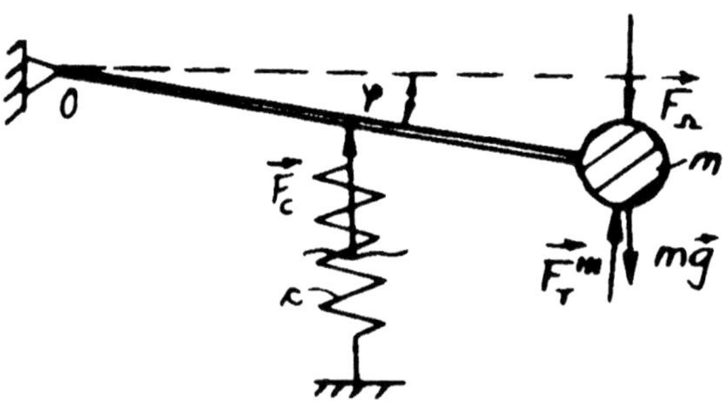


*Figure 5.5.*

Put the condition (5.2.3.) in the last equation:

$$\ddot{\varphi} + \frac{ca^2}{ml^2}\varphi = \frac{F_0}{ml}\sin\Omega t \qquad \text{. . . . . . . . . . (5.2.4.)}$$

5.3. Estimate differential equation for vibration's system in the previous problem (Figure 5.4.) using D'Alembert's law.

Solution:
Add the inertial force for mass *m* and displace the system off the position of equilibrium (Figure 5.5.).

$$\sum M_o = 0$$

$$F_t^{in} l + F_c a\cos\varphi - mgl\cos\varphi - F_\Omega l = 0$$

Tangential inertial force:

$$F_T^{IN} = ma_T = m\frac{dv_m}{dt} = ml\frac{d\dot{\varphi}}{dt} = ml\ddot{\varphi}$$

$$ml^2\ddot{\varphi} + c(f_{st} + a\sin\varphi)a\cos\varphi - mgl\cos\varphi = F_\Omega l$$

Approximate $\sin\varphi \approx \varphi$, and $\cos\varphi \approx 1$ for small vibrations.

$$ml^2\ddot{\varphi} + caf_{st} + ca^2\varphi - mgl = (F_o\sin\Omega t)l \;/ :(ml^2)$$

Put static condition of equilibrium (5.2.3.):

$$\ddot{\varphi} + \frac{ca^2}{ml^2}\varphi = \frac{F_0}{ml}\sin\Omega t$$

The last equation is the same as (4.2.4.).

5.4. Solve previous problem using the law of change of angular momentum.

Solution:
The law of change of angular momentum: $J_0\ddot{\varphi} = M_0^{Fi}$
The mass is concentrated, so individual moment of inertia for the mass equals 0, but inertia's moment of position for the axis in point *O* is product of mass *m* and squared distance $l^2$:

$$J_0 = ml^2$$

The system is swung for the angle $\varphi$ in Figure 5.6. (see the arrow of $\varphi$). Every moment of force in the same sense is positive.

$$ml^2\ddot{\varphi} = -F_c a\cos\varphi + mgl\cos\varphi + F_\Omega l$$

$$ml^2\ddot{\varphi} + c(f_{st} + a\sin\varphi)a\cos\varphi - mgl\cos\varphi = F_\Omega l$$

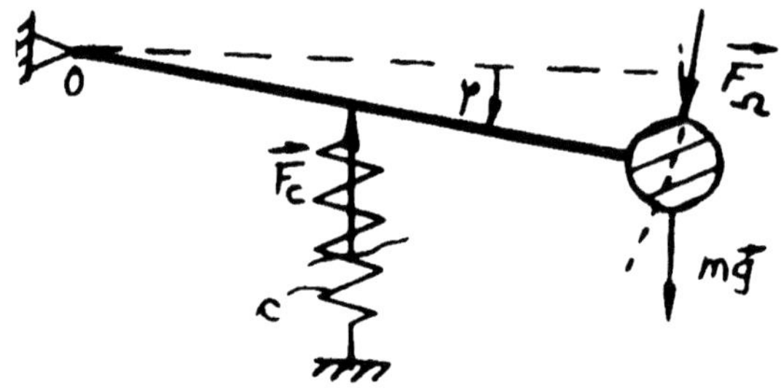


*Figure 5.6.*

Put static condition of equilibrium (5.2.3.), approximation $\sin\varphi \approx \varphi$ and $\cos\varphi \approx 1$ in the last equation:

$$\ddot{\varphi} + \frac{ca^2}{ml^2}\varphi = \frac{F_0}{ml}\sin\Omega t$$

It is equation (5.2.4.).

5.5. Vibration's system is shown in position of equilibrium in Figure 5.7. Calculate physical characteristics of vibrations ($\omega$, $T$) and differential equation, if it is known:
$M$=9[kg], $a$=0,007[m], $c_1$=1500[N/m], $c_2$=1750[N/m], $l = 1[m]$, $l_1 = 0{,}6[m]$, $F_\Omega = a\sin 5t[m, s]$.
The mass $M$ is concentrated, and mass of the lever is neglected.

Final solution:

$$\omega = 21{,}255\left[\frac{rad}{s}\right];$$

$$T = 0{,}296[s];$$

$$\ddot{\varphi} + \left(\frac{c_1 + c_2}{M} - \frac{g}{l}\right)\varphi = \frac{a}{Ml}\sin 5t$$

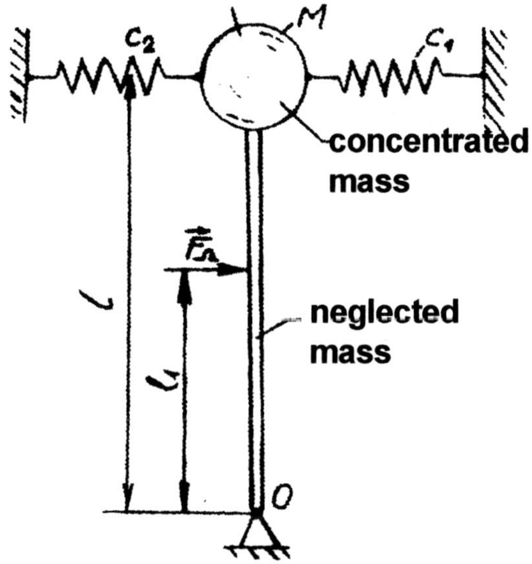


*Figure 5.7.*

5.6. A car is moving with a constant velocity $v_0$ on an uneven road. The form of the road is curve $h = S\sin\frac{2\pi}{L}x$, where *S* is the amplitude and *L* is the length of road's wave. For the mechanical model (Figure 5.8.) estimate the law of excitative motion of a car's box for median road's line (x-axis).

Hint: Use mechanical model in Figure 5.8. for calculation. It is simply shown a car with mass *M* on the springs with rigidity *c* in Figure 5.8.

Final solution:

$$y_p = \frac{\frac{2c}{m}S}{\frac{2c}{m} - \left(\frac{2\pi v_0}{L}\right)^2}\sin\left[\left(\frac{2\pi}{L}v_0\right)t\right]$$

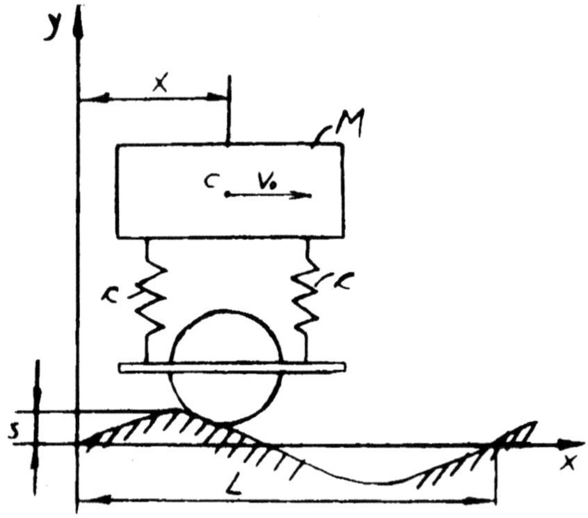


Figure 5.8.

# CHAPTER SIX

## FORCED DAMPED VIBRATIONS WITH ONE DEGREE OF VIBRATION'S FREEDOM

### 6.1. Short Theory Review

There is the resistance to motion like in free damped vibrations (chapter four) and it is necessary to take into consideration the force $F_W$ or function of system's energy dissipation $\phi$. Excitative force $F_\Omega$ acts also upon the system, and it can have form: $F_\Omega = F_0 \cos(\Omega t - \beta)$, like in forced undamped vibrations (chapter five). The general form of differential equation here is:

$$\ddot{q} + 2n\dot{q} + \omega^2 q = h\cos\Omega t \qquad \text{. . . . . . . . . . . . (6.1.)}$$

where $n$ is factor of damping [rad/s], $\omega$ is circular frequency [rad/s], $\Omega$ is frequency of excitative force [rad/s], $q$ is generalized coordinate [rad or m], $t$ is time [s] and $h$ is amplitude of excitation.

General solution of equilibrium (6.1.): $q = q_h + q_p$, where $q_h$ is homogeneous and $q_p$ is particular part of solution.

If $\omega > n$, then

$$q_h = e^{-nt}(C_1 \cos pt + C_2 \sin pt), \qquad \text{. . . . . . (6.2.)}$$

where $p = \sqrt{\omega^2 - n^2}$ is frequency of damped vibrations.

Particular part of the solution has the form:

$$q_p = A\cos(\Omega t - \alpha), \qquad \text{. . . . . . . . . . . . . . . . (6.3.)}$$

where $A$ is dynamic amplitude of forced damped vibrations.

$$A = \frac{h}{\sqrt{(\omega^2 - \Omega^2)^2 + 4n^2\Omega^2}} \qquad \text{. . . . . . . . . . . (6.4.)}$$

$\alpha$ is initial phase of vibrations:

$$\alpha = arctg\frac{2n\Omega}{\omega^2 - \Omega^2} \qquad \text{. . . . . . . . . . . . . (6.5.)}$$

So, the general solution:

$$q = e^{-nt}\left(C_1 \cos\sqrt{\omega^2 - n^2}\,t + C_2 \sin\sqrt{\omega^2 - n^2}\,t\right) + \frac{h}{\sqrt{\left(\omega^2 - \Omega^2\right)^2 + 4n^2\Omega^2}}\cos(\Omega t - \alpha) \quad \ldots(6.6.)$$

Dynamic factor of vibration's increasing (multiplier of resonance):

$$\eta = \frac{A}{A_{st}} = \frac{1}{\sqrt{\left(1-\psi^2\right)^2 + 4\xi^2\psi^2}}, \quad \ldots\ldots(6.7.)$$

where $A_{st} = \frac{h}{\omega^2}$ is static amplitude, $A$ is dynamic amplitude, $\psi = \frac{\Omega}{\omega}$; $\xi = \frac{n}{\omega}$.

LaGrange's equation for forced damped vibrations with one degree of vibration's freedom has form:

$$\frac{d}{dt}\left(\frac{\partial E_k}{\partial \dot{\varphi}}\right) - \frac{\partial E_k}{\partial \varphi} + \frac{\partial \Phi}{\partial \dot{\varphi}} + \frac{\partial E_p}{\partial \varphi} = Q_\Omega \quad \ldots\ldots\ldots(6.8.)$$

where $E_k$ is kinetic, $E_p$ is potential energy of the system, $Q$ is generalized excitative force, and $\phi$ is function of system's energy dissipation.

If there is the law of spring's end motion in problem, like in Figure 6.1. and there isn't the excitative force $F_\Omega$, then there is the kinematic excitation. LaGrange's equation has form:

$$\frac{d}{dt}\left(\frac{\partial E_k}{\partial \dot{\varphi}}\right) - \frac{\partial E_k}{\partial \varphi} + \frac{\partial \Phi}{\partial \dot{\varphi}} + \frac{\partial E_p}{\partial \varphi} = 0 \quad \ldots\ldots(6.9.)$$

The law of spring's end motion influences upon potential energy ($E_p$). It will be explained in problem 6.4.

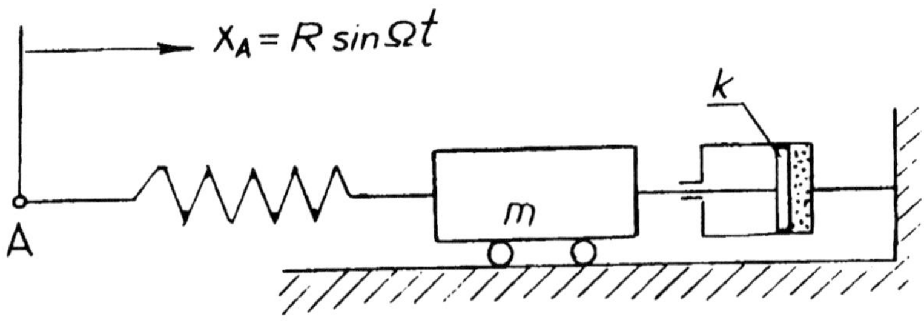


*Figure 6.1.*

## 6.2. Problems

6.1. The homogeneous lever has mass $M$ and length $2l$. It rotates round axis in point $O$. There are two concentrated masses $m$ (Figure 6.2.). The unextensible rope is linked with mass $m$ and thrown over the roller. The roller has radius $r$. A weight is hooked up to the other end of the rope. The weight has mass $m_1$. A muffler has a coefficient of proportionality $k$ and it is linked with mass $m$. Three parallel springs have rigidity $c_1$, $c_2$ and $c_3$. They are linked with upper mass $m$. The excitative force $F_\Omega = 1000 \cdot \sin 5t$ [N,s] acts upon the lever on distance $l/2$ from point $O$.

Estimate:

a) the differential equation of vibrations

b) the law of vibrations, if initial conditions are:

$$\varphi = 0, \quad \dot{\varphi} = 1\frac{m}{s}, \; t_0 = 0$$

It is known:

$M$=$m$=10 [kg], $m_1$=15 [kg], $c_1$=1000 [N/m], $c_2$=1500 [N/m], $c_3$ =500 [N/m], $l$=0,5 [m], $F_\Omega = 1000 \cdot \sin 5t \left[N, s\right]$, $k$=10 [Ns/m].

Solution:

It is necessary to replace three parallel springs with one, which has the equivalent rigidity $c_{ekv}$:

$$c_{ekv} = c_1 + c_2 + c_3 = 1000 + 1500 + 500 = 3000 \quad \left[\frac{N}{m}\right]$$

Static condition of equilibrium (Figure 6.2.):

$$\sum M_0 = 0$$

$$F_{cst} \cdot l - s \cdot l = 0$$

$$c_{ekv} f_{st} \cdot l - s \cdot l = 0$$

$$\sum M_c = 0$$

$$m_1 \cdot g \cdot r - s \cdot r = 0$$

$$s = m_1 \cdot g$$

$$c_{ekv} f_{st} \cdot l - m_1 \cdot g \cdot l = 0 \quad \ldots\ldots\ldots\ldots \quad (6.2.1.)$$

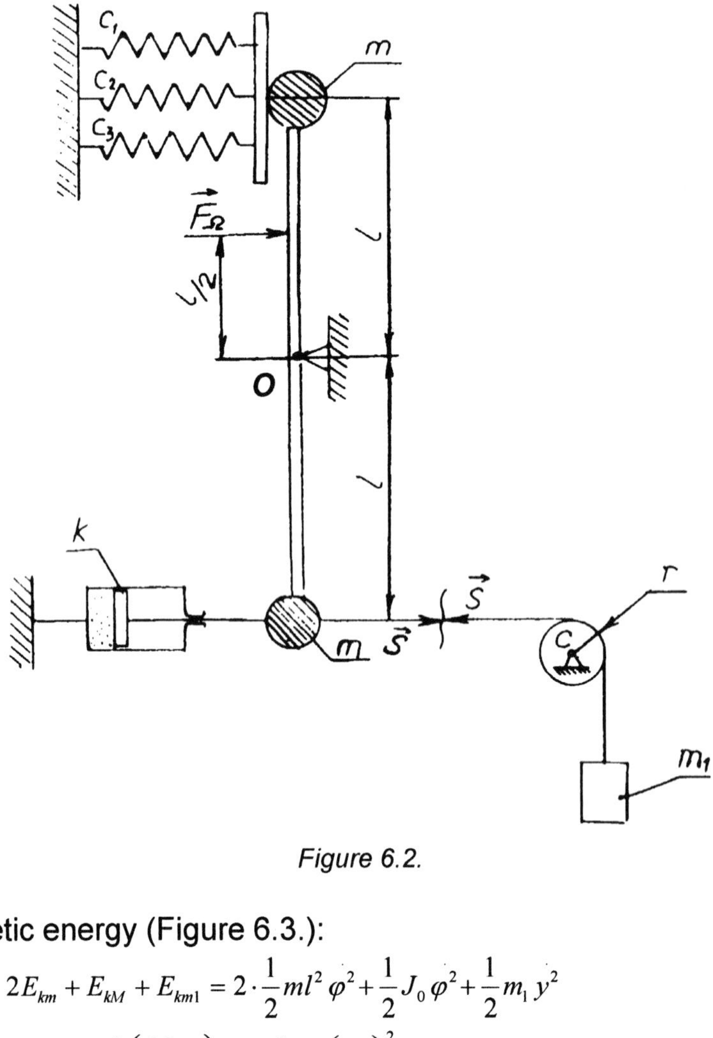


*Figure 6.2.*

Kinetic energy (Figure 6.3.):

$$E_k = 2E_{km} + E_{kM} + E_{km1} = 2 \cdot \frac{1}{2} ml^2 \dot{\varphi}^2 + \frac{1}{2} J_0 \dot{\varphi}^2 + \frac{1}{2} m_1 \dot{y}^2$$

$$E_k = ml^2 \dot{\varphi}^2 + \frac{1}{2}\left(\frac{M}{3} l^2\right)\dot{\varphi}^2 + \frac{1}{2} m_1 \left(l \dot{\varphi}\right)^2$$

The lever's moment of inertia in the last equation (in Figure 6.4.):

$$J_0 = \int_V r^2 \cdot dm = \int_{-l}^{l} x^2 \cdot dm = \rho_1 \int_{-l}^{l} x^2 \cdot dx = \rho_1 \frac{x^3}{3} /_{-l}^{l}$$

$$J_0 = \rho_1 \left(\frac{l^3}{3} - \frac{(-l)^3}{3}\right) = \rho_1 \frac{l^3 + l^3}{3} = \rho_1 \frac{2}{3} l^3$$

$$J_0 = \frac{M}{2l} \cdot \frac{2}{3} l^3 = \frac{Ml}{3},$$

where $\rho_1$ is the mass of length's unit, and $\rho_1 = \frac{M}{2l}$ for the lever. $r=x$, $dm = \rho_1 dx$

It is possible to take the moment of inertia from Table 7. in Appendix.

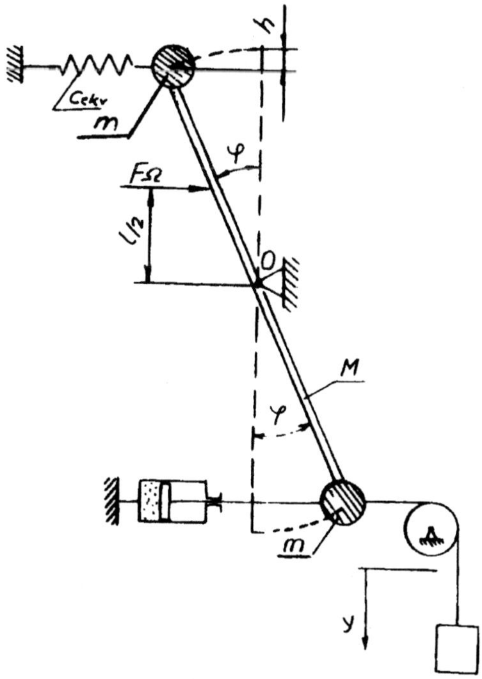


*Figure 6.3.*

Potential energy:

$$E_p = E_{pmg} + E_{pmd} + E_{pm1} + E_{pc}$$

$$E_p = -mgh + mgh - m_1 g \cdot y + \frac{1}{2} c_{ekv} \left(f_{st} + l \sin\varphi\right)^2$$

$$E_p \approx -m_1 g \cdot l\varphi + \frac{1}{2} c_{ekv} \left(f_{st} + l\varphi\right)^2$$

The function of system's energy dissipation $\phi$ (4.14.):

$$\phi = \frac{1}{2} k \cdot v_m^2,$$

where $v_m^2$ is the velocity of mass $m$, and:

$$v_m^2 = l \cdot \dot{\varphi}^2$$

$$\phi = \frac{1}{2} k \cdot v_m^2 = \frac{1}{2} k \cdot l^2 \cdot \dot{\varphi}^2$$

LaGrange's equation:

$$\frac{d}{dt}\left(\frac{\partial E_k}{\partial \dot{\varphi}}\right) - \frac{\partial E_k}{\partial \varphi} + \frac{\partial \Phi}{\partial \dot{\varphi}} + \frac{\partial E_p}{\partial \varphi} = F_\Omega \cdot \frac{l}{2}$$

$$2ml^2 \ddot{\varphi} + 2\frac{1}{6}Ml^2 \ddot{\varphi} + 2\frac{1}{2}ml^2 \ddot{\varphi} - m_1 gl + \frac{1}{2}c_{ekv} \cdot 2(f_{st} + l\varphi) \cdot l +$$

$$+ 2\frac{1}{2}kl^2 \dot{\varphi} = 1000\frac{l}{2}\sin 5t$$

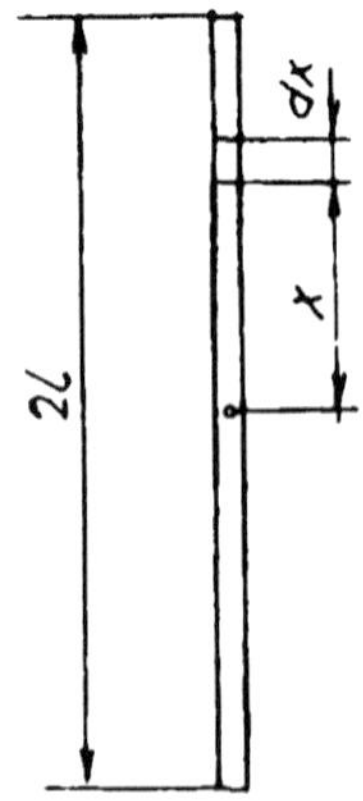

*Figure 6.4.*

Put the static condition (6.2.1.) in the last equation and set it in order:

$$\ddot{\varphi} + \frac{k}{2m + \frac{M}{3} + m_1} \cdot \dot{\varphi} + \frac{c_{ekv}}{2m + \frac{M}{3} + m_1} \cdot \varphi = \frac{1000}{2l(2m + \frac{M}{3} + m_1)} \sin 5t$$

The homogeneous part of the solution has this form:

$$\varphi_H = e^{-nt}\left[C_1 \cos\left(\sqrt{\omega^2 - n^2} \cdot t\right) + C_2 \sin\left(\sqrt{\omega^2 - n^2} \cdot t\right)\right]$$

Derive the last equation with the time *t*:

$$\dot{\varphi}_H = -n e^{-nt}(C_1 \cos pt + C_2 \sin pt) +$$

$$+ e^{-nt}(-C_1 p \sin pt + C_2 p \cos pt)$$

where $p = \sqrt{\omega^2 - n^2}$.

When we put the initial conditions in the equations for $\varphi_H$ and $\dot{\varphi}_H$:

$$0 = 1 \cdot (C_1 + 0) \quad \Rightarrow \quad C_1 = 0$$

$$1 = -n \cdot 1(C_1 + 0) + 1 \cdot (0 + C_2 \cdot p) \quad \Rightarrow \quad C_2 = \frac{1}{p}\left[\frac{\frac{rad}{s}}{\frac{1}{s}} = rad\right]$$

Compare the last differential equation with its general form (6.1.):

$$\omega^2 = \frac{c_{ekv}}{2m + \frac{M}{3} + m_1} = \frac{3000\left[\frac{N}{m}\right]}{2 \cdot 10[kg] + \frac{10}{3}[kg] + 15[kg]} = 78{,}26\left[s^{-2}\right]$$

$$\omega = 8{,}85\left[s^{-1}\right]$$

$$2n = \frac{k}{2m + \frac{M}{3} + m_1} = \frac{10\left[\frac{Ns}{m}\right]}{2 \cdot 10[kg] + \frac{10}{3}[kg] + 15[kg]} = 0{,}26\left[s^{-1}\right]$$

$$n = 0{,}13\left[s^{-1}\right]$$

$$p = \sqrt{78{,}26\left[s^{-2}\right] - 0{,}13^2\left[s^{-2}\right]} = 8{,}85\left[s^{-1}\right]$$

$$C_2 = \frac{1}{8{,}85} = 0{,}113[rad]$$

$$\varphi_H = e^{-0{,}13t} \cdot 0{,}113 \sin 8{,}85t\,[rad, s]$$

It is needed to suppose the particular part of this solution in the form of excitative force:

$$\varphi_p = A \sin \Omega t \Rightarrow \qquad \varphi_p = \frac{h}{\sqrt{\left(\omega^2 - \Omega^2\right)^2 + 4n^2\Omega^2}} \sin \Omega t$$

$$h = \frac{1000}{2l \cdot \left(2m + \frac{M}{3} + m_1\right)} =$$

$$= h = \frac{1000[N]}{2 \cdot 0{,}5[m] \cdot \left(2 \cdot 10[kg] + \frac{10}{3}[kg] + 15[kg]\right)} = 26{,}09\left[s^{-2}\right]$$

$$\varphi_p = \frac{26{,}09\left[s^{-2}\right]}{\sqrt{\left(8{,}85^2 - 5^2\right)^2\left[s^{-4}\right] + 4\cdot 0{,}13^2\left[s^{-2}\right]\cdot 5^2\left[s^{-2}\right]}} \sin 5\left[s^{-1}\right]t$$

$$\varphi_p = 0{,}49\sin 5t\left[rad, s\right]$$

$$\varphi = \varphi_H + \varphi_p$$

6.2. A lever is hung on point *O* and the excitative force acts upon it (Figure 6.5.). The muffler is situated on the end of the lever. The lever is linked for immovable wall with the spring. The muffler has coefficient *k* and the spring has rigidity *c*. Calculate the period of vibrations *T*, amplitude of excitation *h* and damping factor *n*, using the law of change of angular momentum, if it is known: *k, M, l, c* and $F_\Omega = F_O \sin\Omega t$.

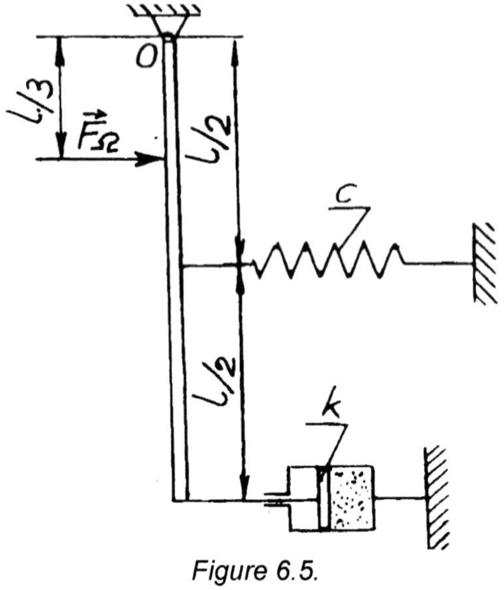


*Figure 6.5.*

Solution:

The law of change of angular momentum (Figure 6.6.):

$$J_o \cdot \ddot{\varphi} = \sum M_o^{Fi}$$

$$\frac{1}{3}Ml^2 \cdot \ddot{\varphi} = F_\Omega \cdot \frac{l}{3}\cos\varphi - F_c \cdot \frac{l}{2}\cos\varphi - F_W \cdot l\cos\varphi - Mg \cdot \frac{l}{2} \cdot \sin\varphi$$

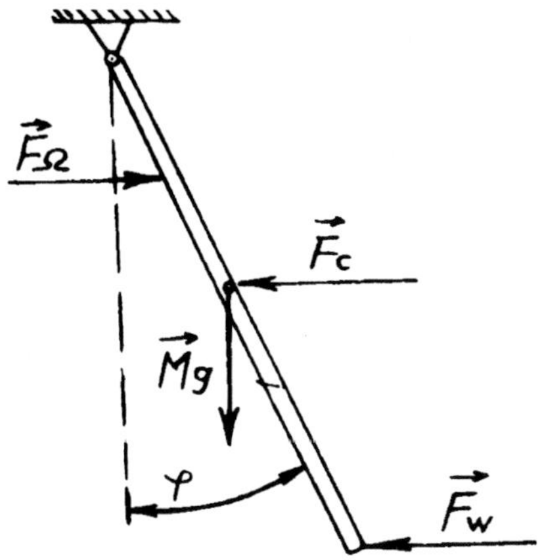


*Figure 6.6.*

For small vibrations: cos$\varphi\approx$1 and sin$\varphi\approx\varphi$

$$\frac{1}{3}Ml^2\cdot\ddot{\varphi}=F_\Omega\cdot\frac{l}{3}-F_c\cdot\frac{l}{2}-F_W\cdot l-Mg\cdot\frac{l}{2}\cdot\varphi$$

$$\frac{1}{3}Ml^2\cdot\ddot{\varphi}=F_o\cdot\frac{l}{3}\sin\Omega t-c\cdot\frac{l^2}{4}\cdot\varphi-k\cdot l^2\cdot\dot{\varphi}-Mg\cdot\frac{l}{2}\cdot\varphi$$

$$\frac{1}{3}Ml^2\cdot\ddot{\varphi}+k\cdot l^2\cdot\dot{\varphi}+c\cdot\frac{l^2}{4}\cdot\varphi+Mg\cdot\frac{l}{2}\cdot\varphi=F_o\cdot\frac{l}{3}\sin\Omega t$$

$$\ddot{\varphi}+\frac{3k}{M}\dot{\varphi}+\left(\frac{3}{4}\cdot\frac{c}{M}+\frac{3}{2}\cdot\frac{g}{l}\right)\cdot\varphi=\frac{F_o}{Ml}\sin\Omega t \qquad \text{........ (6.2.2.)}$$

Compare equations (6.2.2.) and (6.1.):

$$\omega^2=\frac{3}{4}\cdot\frac{c}{M}+\frac{3}{2}\cdot\frac{g}{l} \quad\Rightarrow\quad \omega=\sqrt{\frac{3}{4}\cdot\frac{c}{M}+\frac{3}{2}\cdot\frac{g}{l}}$$

$$T=\frac{2\pi}{\omega}=\frac{2\pi}{\sqrt{\frac{3}{4}\cdot\frac{c}{M}+\frac{3}{2}\cdot\frac{g}{l}}}$$

$$h=\frac{F_o}{Ml};\quad 2n=\frac{3k}{M}\quad\Rightarrow\quad n=\frac{3}{2}\cdot\frac{k}{M}$$

6.3. Solve the previous problem using LaGrange's equation.

Solution:

It is necessary to write kinetic ($E_k$), potential ($E_p$) energy and the function of system's energy dissipation ($\phi$).

$$E_k = \frac{1}{2} J_0 \cdot \dot{\varphi}^2 = \frac{1}{2} \cdot \frac{1}{3} Ml^2 \cdot \dot{\varphi}^2$$

$$E_p = \left( \frac{l}{2} - \frac{l}{2} \cos \varphi \right) \cdot Mg + \frac{1}{2} c \cdot \left( \frac{l}{2} \sin \varphi \right)^2 \approx$$

$$\approx Mg \frac{l}{2} \left( 1 - 1 + \frac{\varphi^2}{2} \right) + \frac{1}{2} c \frac{l^2}{4} \dot{\varphi}^2$$

$$E_p = \frac{1}{4} Mgl \cdot \varphi^2 + \frac{1}{8} cl^2 \cdot \varphi^2$$

$$\Phi = \frac{1}{2} kl^2 \cdot \dot{\varphi}^2$$

There is the moment of force $F_\Omega$, on the right side of equation (6.8.):

$$F_\Omega \cdot \frac{l}{3} = F_o \cdot \frac{l}{3} \cdot \sin \Omega t$$

$$\frac{d}{dt}\left( \frac{\partial E_k}{\partial \dot{\varphi}} \right) - \frac{\partial E_k}{\partial \varphi} + \frac{\partial \Phi}{\partial \dot{\varphi}} + \frac{\partial E_p}{\partial \varphi} = Fo \cdot \frac{l}{3} \sin \Omega t$$

$$2Ml^2 \ddot{\varphi} + kl^2 \dot{\varphi} + \frac{1}{2} Mgl\varphi + \frac{1}{4} cl^2 \cdot \varphi = F_o \cdot \frac{l}{3} \sin \Omega t$$

$$\ddot{\varphi} + \frac{3k}{M} \dot{\varphi} + \left( \frac{3}{2} \cdot \frac{g}{l} + \frac{3}{4} \cdot \frac{c}{M} \right) \cdot \varphi = \frac{F_o}{Ml} \sin \Omega t$$

The last equation is equivalent to (6.2.2.).

6.4. It is necessary to determine the differential equation of vibrations for the same vibration's system from problem 6.1, if the excitative force $F_\Omega$ doesn't act upon it, but the left end of the spring moves, forced by the law $X_A = 0{,}01 \sin 5t$ like in figure 6.7.

Solution:

We can use LaGrange's equation:

$$E_k = \frac{1}{2} \cdot 2ml^2 \cdot \dot{\varphi}^2 + \frac{1}{6} Ml^2 \cdot \dot{\varphi}^2 + \frac{1}{2} m_1 l^2 \cdot \dot{\varphi}^2$$

$$E_p = -m_1 g l \varphi + \frac{1}{2} c_{ekv} \left( f_{st} + l\varphi - x_A \right)^2$$

$$\Phi = \frac{1}{2} k \cdot l^2 \cdot \dot{\varphi}^2$$

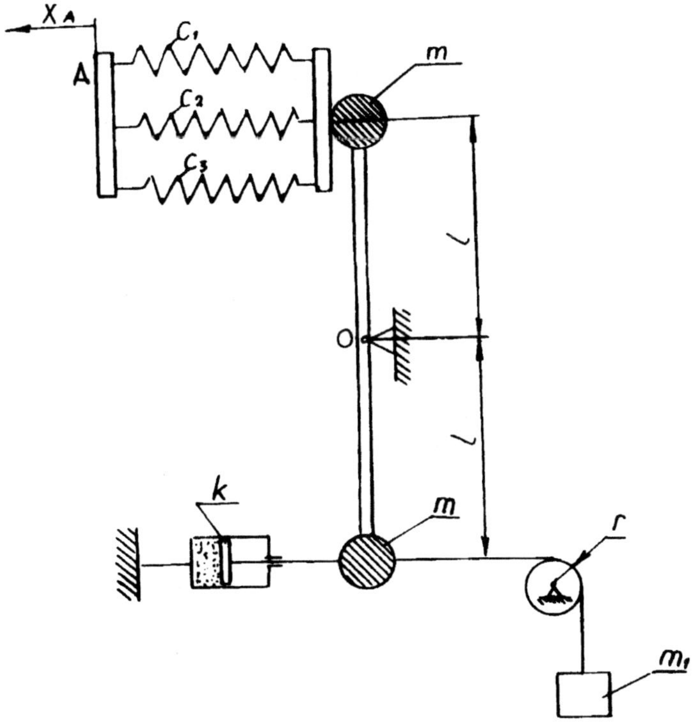


*Figure 6.7.*

It is needed to use LaGrange's equation in form (6.9.), because it gives the law of motion for spring's end:

$$\frac{d}{dt}\left( \frac{\partial E_k}{\partial \dot{\varphi}} \right) - \frac{\partial E_k}{\partial \varphi} + \frac{\partial \Phi}{\partial \dot{\varphi}} + \frac{\partial E_p}{\partial \varphi} = 0$$

$$2ml^2 \ddot{\varphi} + \frac{1}{3} Ml^2 \ddot{\varphi} + m_1 l^2 \ddot{\varphi} - m_1 g l + c_{ekv} \left( f_{st} + l\varphi - X_A \right) \cdot l + kl^2 \dot{\varphi} = 0$$

$$\left( 2ml^2 + \frac{1}{3} Ml^2 + m_1 l^2 \right) \cdot \ddot{\varphi} + kl^2 \cdot \dot{\varphi} + c_{ekv} l^2 \cdot \varphi = c_{ekv} l X_A + m_1 g l - c_{ekv} l f_{st}$$

Static equilibrium is (6.2.1.):

$$\left(2ml^2+\frac{1}{3}Ml^2\right)\cdot\ddot{\varphi}+kl^2\cdot\dot{\varphi}+c_{ekv}l^2\cdot\varphi=c_{ekv}l\cdot 0{,}01\sin 5t$$

$$\ddot{\varphi}+\frac{k}{2m+\frac{M}{3}+m_1}\dot{\varphi}+\frac{c_{ekv}}{2m+\frac{M}{3}+m_1}\varphi=\frac{c_{ekv}\cdot 0{,}01}{\left(2m+\frac{M}{3}+m_1\right)l}\sin 5t$$

6.5. Excitative force $F_\Omega$ acts upon concentrated mass $m$. The mass is linked to the ceiling by two springs and to the ground by two mufflers. Rigidities of the springs are $c_1$ and $c_2$. The mufflers have coefficients $k_1$ and $k_2$. It is given:
$k_1$=30 [Ns/m]; $k_2$=10 [Ns/m]; $F_\Omega$=1,7sin2t [N,s]; $c_1$=3000 [N/m]; $c_2$=1500 [N/m]; $m$=2 [kg].
Estimate:
a) the circular frequency $\omega$, the factor of damping $n$ and the amplitude of excitation $h$
b) the law of vibrations, if the initial conditions are: $t_o=0[s]$, $(y_H)_o=0{,}02[m]$, $(\dot{y}_H)_o=0[m/s]$

Solution:

$$E_k=\frac{1}{2}m\cdot\dot{y}^2$$

$$\Phi=\frac{1}{2}k_{ekv}\cdot\dot{y}^2; k_{ekv}=k_1+k_2=30+10=40\left[\frac{Ns}{m}\right]$$

$$E_p=\frac{1}{2}c_{ekv}(y+y_{st})^2-mg\cdot y;$$

$$\frac{1}{c_{ekv}}=\frac{1}{c_1}+\frac{1}{c_2}=\frac{1}{30}+\frac{1}{15}=\frac{1}{10}\quad\Rightarrow\quad c_{ekv}=10\left[\frac{N}{cm}\right]$$

$$\frac{d}{dt}\left(\frac{\partial E_k}{\partial\dot{y}}\right)-\frac{\partial E_k}{\partial y}+\frac{\partial\Phi}{\partial\dot{y}}+\frac{\partial E_p}{\partial y}=F_\Omega$$

$$m\ddot{y}+k_{ekv}\dot{y}+c_{ekv}(y+y_{st})-mg=1{,}7\sin 2t \quad\ldots\ldots (6.2.3.)$$

The static condition of equilibrium (Figure 6.8.):

$$mg-c_{ekv}\cdot y_{st}=0$$

Put it in the equation (6.2.3.):

$$m\ddot{y}+k_{ekv}\dot{y}+c_{ekv}y=1{,}7\sin 2t$$

$$\ddot{y}+\frac{k_{ekv}}{m}\cdot\dot{y}+\frac{c_{ekv}}{m}\cdot y=\frac{1{,}7}{m}\sin 2t$$

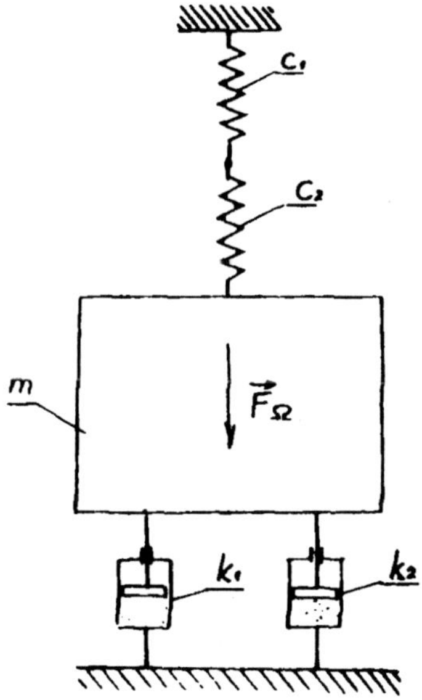


Figure 6.8.

a)

$$2n=\frac{c_{ekv}}{m}\Rightarrow n=\frac{k_{ekv}}{2m}=\frac{40\left[\frac{Ns}{m}\right]}{2\cdot 2[kg]}=10\left[s^{-1}\right]$$

$$\omega^2=\frac{c_{ekv}}{m}=\frac{1000\left[\frac{N}{m}\right]}{2\geq [kg]}=500\left[s^{-2}\right]\;\Rightarrow\;\omega=22{,}36\left[s^{-1}\right]$$

So, this is the case of low damping $\omega > n$

$$h=\frac{1{,}7}{m}=\frac{1{,}7}{2}=0{,}85\left[\frac{m}{s^2}\right]$$

b)

$$y_H=e^{-nt}\left(C_1\cos pt+C_2\sin pt\right);$$

$$p=\sqrt{\omega^2-n^2}=\sqrt{500-100}=20\left[s^{-1}\right]$$

$$\dot{y}_H = e^{-nt} \cdot (-n)(C_1 \cos pt + C_2 \sin pt) +$$
$$+ e^{-nt}(-C_1 p \sin pt + C_2 p \cos pt)$$

Put the initial conditions in equations for $y_H$ and $\dot{y}_H$:

$$0{,}02 = 1 \cdot (C_1 \cdot 1 + 0) \quad \Rightarrow \quad C_1 = 0{,}02[m]$$
$$0 = 1 \cdot (-10)(C_1 + 0) + 1 \cdot (0 + C_2 \cdot 20 \cdot 1)$$
$$20C_2 = 10C_1 \quad \Rightarrow \quad C_2 = \frac{0{,}02 \cdot 10}{20} = 0{,}01[m]$$
$$y_H = e^{-10t}(0{,}02 \cos 20t + 0{,}01 \sin 20t)[m, s]$$

The particular part of solution:

$$y_p = \frac{h}{\sqrt{(\omega^2 - \Omega^2)^2 + 4n^2\Omega^2}} \sin \Omega t$$

$$y_p = \frac{0{,}85\left[\frac{m}{s^2}\right]}{\sqrt{\left(500\left[s^{-2}\right] - 4\left[s^{-2}\right]\right)^2 + 4 \cdot 100\left[s^{-2}\right] \cdot 4\left[s^{-2}\right]}} \sin 2\left[s^{-2}\right]$$

$$y_p = 0{,}0017 \sin 2t[m, s]$$

The general solution: $\qquad y = y_H + y_p$

6.6. The device for testing vertical vibrations is shown in Figure 6.9. The motor *A* and the stand *B* have weight *G* together. The motor and stand are linked to the ground by two springs and a muffler. The periodical force: $F_\Omega = F_0 \cdot \sin \Omega t$ acts upon the motor. $F_0$ is the amplitude of excitative force $F_\Omega$ and $\Omega$ is the angular speed of rotary motor's parts, i.e. $\Omega$ is the frequency of $F_\Omega$. It is known: $G$=98,1[N]; $k$=100[Ns/m]; $c$=2000[N/m]; $F_0$=1200[N]; $\Omega$=10[$s^{-1}$].

Estimate:

a) the excitative motion of the motor ($y_p$),

b) the rigidity $c$ of the springs, if the dynamic factor of vibration's increasing $\eta$=0,25

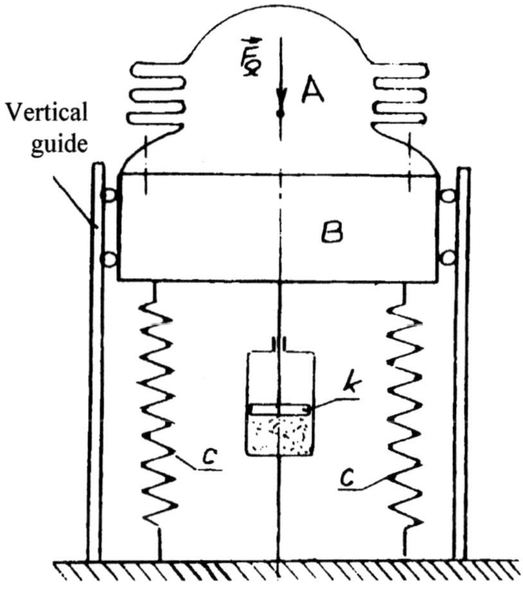


*Figure 6.9.*

Solution:

Newton's second law (Figure 6.10):

$$\frac{G}{g}\cdot\ddot{y}=\sum F_{iy}$$

$$\frac{G}{g}\cdot\ddot{y}=F_{\Omega}-F_{c_{ekv}}-F_{W}$$

$$F_{c_{ekv}}=2cy;\quad F_{W}=k\dot{y}$$

$$\frac{G}{g}\cdot\ddot{y}+k\dot{y}+2cy=F_0\sin\Omega t$$

$$\ddot{y}+\frac{k\cdot g}{G}\dot{y}+\frac{2c\cdot g}{G}y=\frac{F_0\cdot g}{G}\sin\Omega t \qquad \text{. . . . (6.2.4.)}$$

a) A particular part of the solution can have form (6.3.), i.e. (2.6.) (the constants are *A* and $\alpha$) or form (2.3.) (the constants are $C_1$ and $C_2$). It is possible to write form (2.3.) like:

$$y_p=M\sin\Omega t+N\cos\Omega t$$

The constants are *M* and *N*, and it is necessary to calculate them:

$$\dot{y}_p = M\Omega \cos \Omega t - N\Omega \sin \Omega t$$

$$\ddot{y}_p = -M\Omega^2 \sin \Omega t - N\Omega^2 \cos \Omega t$$

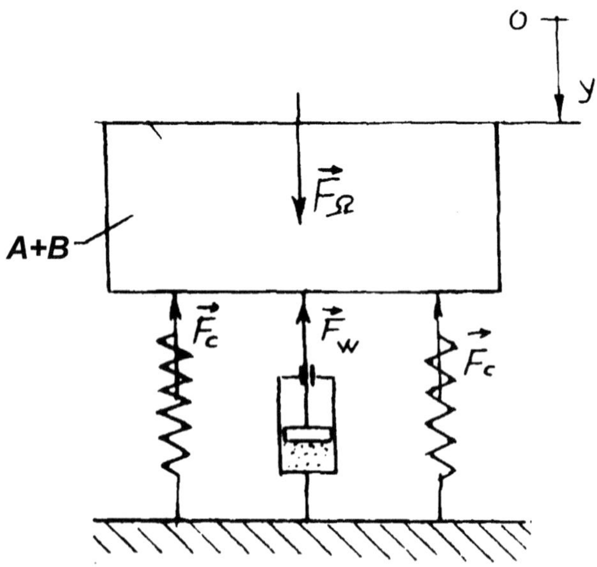


*Figure 6.10.*

Put $y_p$, $\dot{y}_p$ and $\ddot{y}_p$ in equation (6.2.4.):

$$\sin \Omega t\left(- M\Omega^2 - 100N\Omega + 400M\right) +$$
$$+ \cos \Omega t\left(- N\Omega^2 + 100M\Omega + 400N\right) = 120 \cdot \sin \Omega t$$

The coefficient beside sin$\Omega t$ in the left side is equal to the coefficient beside sin$\Omega t$ in the right side. The same for cos$\Omega t$. So, we give the system two equations:

$$400M - 100M - 100N = 120$$
$$400N + 100M - 100N = 0$$

$$N = -\frac{1}{3}M$$

$$300M + 100 \cdot \frac{1}{3}M = 120 \Rightarrow M = 0{,}36[m]$$

$$N=-\frac{1}{3}M=-\frac{1}{3}\cdot 0{,}36=-0{,}12[m]$$

The excitative motion of the motor:

$$y_p=0{,}36\sin 10t-0{,}12\cos 10t\ [m,s]$$

b)

$$\eta=\frac{1}{\sqrt{\left(1-\psi^2\right)^2+4\xi^2\psi^2}};\qquad \xi=\frac{n}{\omega}=\frac{5\left[s^{-1}\right]}{20\left[s^{-1}\right]}=0{,}25;\qquad \psi=\frac{\Omega}{\omega}$$

Square the expression of $\eta$:

$$\left(1-\psi^2\right)^2+4\xi^2\psi^2=\frac{1}{\eta^2}$$

$$1-2\psi^2+\psi^4+4\xi^2\psi^2=\frac{1}{\eta^2}$$

$$\psi^4-\left(4\xi^2-2\right)\cdot\psi^2-\frac{1}{\eta^2}=0 \qquad \text{.........(6.2.5.)}$$

Put $\xi$=0,25 and $\eta$=0,25 in equation (6.2.5.):

$$\psi^4+1{,}75\cdot\psi^2-16=0$$

It is a biquadratic equation. Bring in the substitution: $\psi^2=s$

$$s^2+1{,}75\cdot s-16=0$$

$$s_{1,2}=\frac{-1{,}75\pm 8{,}19}{2},\quad \text{so}\quad s_1=-4{,}97\quad \text{and}\quad s_2=3{,}22$$

$$\text{i.e.:}\quad \psi_1^2=-4{,}97\quad \text{and}\quad \psi_2^2=3{,}22$$

Negative solution ($\psi_1^2$) is senseless, because $\psi^2$ can't be negative, so the solution is $\psi_2^2$=3,22.

From equation (6.2.4.): $\omega=\sqrt{\frac{2cg}{G}}$,

The rigidity *c* of the springs from the last equation:

$$\psi=\frac{\Omega}{\omega}=\frac{\Omega}{\sqrt{\frac{2cg}{G}}}\quad \Leftrightarrow\quad \frac{2cg}{G}=\frac{\Omega^2}{\psi^2}$$

$$c = \frac{G \cdot \Omega^2}{\psi^2 \cdot 2 \cdot g} = \frac{98{,}1[N] \cdot 10^2 \left[s^{-2}\right]}{3{,}22^2 \cdot 2 \cdot 9{,}81 \left[\frac{m}{s^2}\right]} = 155{,}28 \left[\frac{N}{m}\right]$$

6.7. The mechanical model of a vibrational machine is shown in Figure 6.11. Mass of the machine is *m*. Eccentric masses $m_1$ rotate in an inverse sense with angular speed $\Omega$ and make the excitative force $F_\Omega$. All resistance to motion is taken into consideration with the muffler, which has the coefficient of proportion *k*.

Estimate:

a) the differential equation of vibrations, factor of damping *n*, circular frequency and amplitude of excitation *h*, if it is known: $m$=5$m_1$=5 [kg]; $c$=2000 [N/m]; $k$=10 [Ns/m]; $r$=0,05 [m]; $\Omega$=30 [$s^{-1}$],

b) Critical number of revolutions $n_{kr}$ of eccentric masses (It is the number of revolutions when the resonance appears)

Solution:

a) Newton's Second law (Figure 5.12.)

$$m \cdot \ddot{x} = \sum F_{ix} t$$

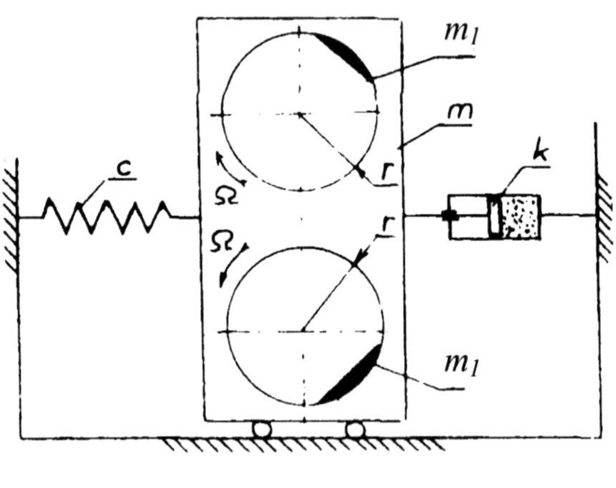


*Figure 6.11.*

$m \cdot \ddot{x} = 2F_{\Omega} - F_W - F_c$

$m \cdot \ddot{x} = 2F_{\Omega} - k \cdot \dot{x} - cx$

$m \cdot \ddot{x} + k \cdot \dot{x} + cx = 2F_m \cos\varphi$

$$\ddot{x} + \frac{k}{m} \cdot \dot{x} + \frac{c}{m} \cdot x = 2 \cdot \frac{F_m}{m} \cos \Omega t$$

The centrifugal force: $F_m = m_1 \cdot a_n$, where $a_n$ is the normal acceleration: $a_n = \cdot r \cdot \Omega^2$,

$F_m = m_1 \cdot r \cdot \Omega^2$

$$\ddot{x} + \frac{k}{m} \cdot \dot{x} + \frac{c}{m} \cdot x = 2 \cdot \frac{1}{5} \cdot r \cdot \Omega^2 \cdot \cos \Omega t$$

$$2n = \frac{k}{m} \Rightarrow n = \frac{k}{2 \cdot m} = \frac{10[Ns/m]}{2 \cdot 5[kg]} = 1\left[s^{-1}\right]$$

$$\omega = \sqrt{\frac{c}{m}} = \sqrt{\frac{2000[N/m]}{5[kg]}} = 20\left[s^{-1}\right];$$

$$h = 2 \cdot \frac{1}{5} \cdot r \cdot \Omega^2 = 2 \cdot \frac{1}{5} \cdot 0{,}05[m] \cdot 30^2\left[s^{-2}\right] = 18\left[\frac{m}{s^2}\right]$$

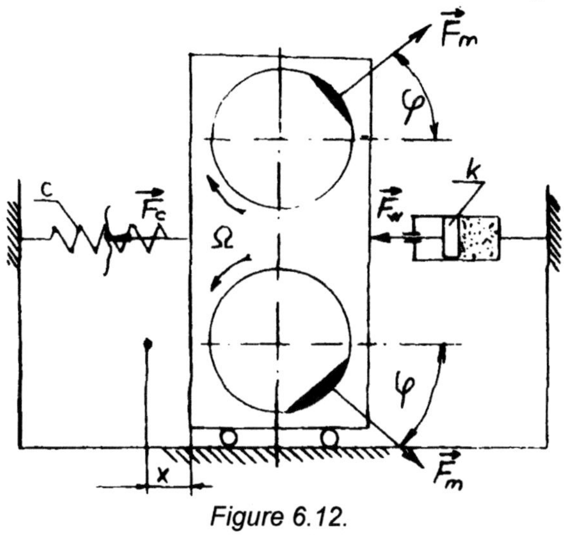


*Figure 6.12.*

b) *n* is the number of revolutions for eccentric masses:

$$\Omega = \frac{2 \cdot n \cdot \pi}{60} \quad \Rightarrow \quad n = \frac{60 \cdot \Omega}{2\pi} = \frac{30 \cdot \Omega}{\pi}$$

The resonance will appear when $\Omega = \omega$, and then $n = n_{kr}$:

$$n_{kr} = \frac{30}{\pi} \cdot \omega = \frac{30}{\pi} \cdot \sqrt{\frac{c}{m}} = \frac{30}{\pi} \cdot \sqrt{\frac{2000\left[\frac{N}{m}\right]}{5[kg]}} = 190{,}9 \quad \left[{}^{o}\!/_{\min}\right]$$

6.8. The plate in Figure 6.13. has weight $G$=0,5[kN] and it is hung with the spring. The rigidity of the spring is $c$=0,25[KN/cm]. The plate is in fluid. The other end of the spring is linked with slider $A$, which does vertical vibrations by law: $y = b \cdot \sin \Omega t$, where $b$=2 [cm], and $\Omega = 15$[rad/s]. The force of liquid's resistance $F_w$ acts upon the plate and $F_w = 3{,}05$ [mN].

Estimate the excitative motion of the plate (a particular part of the solution) when the plate's speed is 1[cm/s].

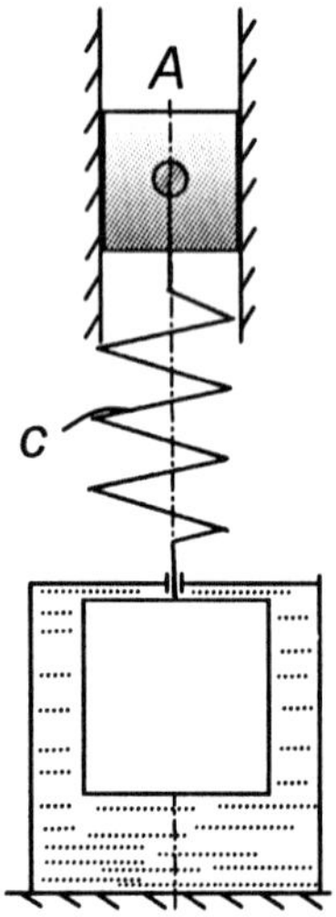


*Figure 6.13.*

Final solution:

$$y_p = 3{,}5\sin(15t - 0{,}3274)\,[cm, s]$$

6.9. The machine in Figure 6.14. has weight $G$=100[N]. There are two eccentric masses (item 1) and they rotate in an inverse sense. The weight of eccentric masses is $G_1$=1[N], and the eccentricity is $r$=2[cm]. Rigidity of the spring is $c$=100[N/cm] and the factor of damping $n$=2[s$^{-1}$]. Estimate the critical number of revolutions $n_{kr}$ of the eccentric masses, when the resonance appears. How much will the dynamic amplitude $A_{din}$ and dynamic factor of vibration's increasing $\eta_{rez}$ be at the moment of resonance?

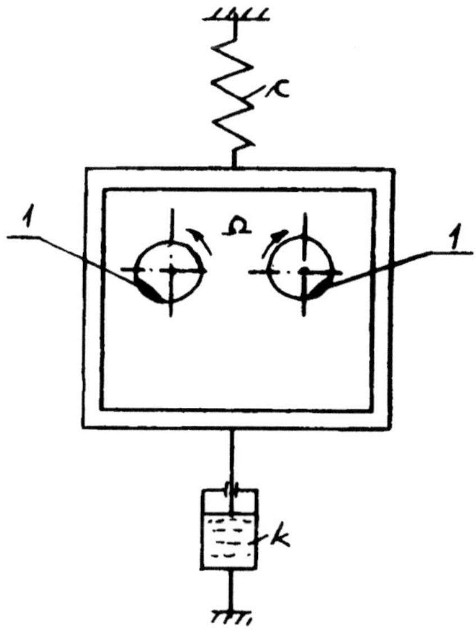


*Figure 6.14.*

Final solution:

$n_{kr}$=300[$^r$/min], $A_{din}$=0,314[cm], $\eta_{rez} = 7{,}85$

# CHAPTER SEVEN

## FREE UNDAMPED VIBRATIONS WITH TWO DEGREES OF VIBRATION'S FREEDOM

### 7.1. Short Theory Review

The best way for solving the problems in this part is using LaGrange's equations of second order. Here they have form:

$$\begin{aligned} &\frac{d}{dt}\left(\frac{\partial E_k}{\partial \dot{q}_1}\right)-\frac{\partial E_k}{\partial q_1}+\frac{\partial E_p}{\partial q_1}=0 \\ &\frac{d}{dt}\left(\frac{\partial E_k}{\partial \dot{q}_2}\right)-\frac{\partial E_k}{\partial q_2}+\frac{\partial E_p}{\partial q_2}=0 \end{aligned} \quad ,\ .\ .\ .\ .\ .\ .\ .\ .\ .\ (7.1.)$$

where $q_1$ and $q_2$ are generalized coordinates. The coordinates determine the position of the vibration's system.

The kinetic energy $E_k$:

$$E_k=\frac{1}{2}\left(a_{11}\dot{q}_1^2+2a_{12}\dot{q}_1\dot{q}_2+a_{22}\dot{q}_2^2\right),$$

where $a_{11}$; $a_{12}=a_{21}$; $a_{22}$ - are the coefficients of system's inertia.

The potential energy $E_p$:

$$E_p=\frac{1}{2}\left(c_{11}q_1^2+2c_{12}q_1q_2+c_{22}q_2^2\right),$$

where $c_{11}$; $c_{12}=c_{21}$; $c_{22}$ - are the coefficients of system's rigidity.

The coefficients $a_{ij}$ and $c_{ij}$ are the parameters of vibration's system - masses rigidities etc. Put $E_k$ and $E_p$ in the system (7.1.):

$$\begin{aligned} &a_{11}\ddot{q}_1+a_{12}\ddot{q}_2+c_{11}q_1+c_{12}q_2=0 \\ &a_{21}\ddot{q}_1+a_{22}\ddot{q}_2+c_{21}q_1+c_{22}q_2=0 \end{aligned} \quad ,\ .\ .\ .\ .\ .\ .(7.2.)$$

It is the general form of the differential equations system.

The equations (7.2.) are system of homogeneous differential equations of second order with constant coefficient, and partial solutions are:

$$\begin{aligned} q_1 &= A_1 \sin(\omega t + \beta) \\ q_2 &= A_2 \sin(\omega t + \beta) \end{aligned} \quad \text{. . . . . . . . . . . . . . . (7.3.)}$$

where$\omega$ is the circular frequency, $\beta$ is the initial phase of vibrations.

If we put (7.3.) in system (7.2.), we will get the system of algebraic equations:

$$\begin{aligned} A_1 \cdot \left(-a_{11}\omega^2 + c_{11}\right) + A_2 \cdot \left(-a_{12}\omega^2 + c_{12}\right) = 0 \\ A_1 \cdot \left(c_{12} - a_{12} \cdot \omega^2\right) + A_2 \cdot \left(c_{22} - a_{22}\omega^2\right) = 0 \end{aligned} \quad \text{. . . . .(7.4.)}$$

The solutions of system (7.4.) won't be trivial, if determinant of the system is equal to 0.

$$\Delta\left(\omega^2\right) = \begin{vmatrix} c_{11} - a_{11}\omega^2 & c_{12} - a_{12}\omega^2 \\ c_{12} - a_{12}\omega^2 & c_{22} - a_{22}\omega^2 \end{vmatrix} = 0 \quad \text{. . . . . (7.5.)}$$

We will get the frequent equation with solving the determinant.

$$\left(c_{11} - a_{11}\omega^2\right) \cdot \left(c_{22} - a_{22}\omega^2\right) - \left(c_{12} - a_{12}\omega^2\right)^2 = 0 \quad \text{. . . . (7.6.)}$$

The solutions of the frequent equation (7.6.) are $\omega^2{}_1$ and $\omega^2{}_2$ ($\omega_1 < \omega_2$).

For $\omega_1$ we have the first main form of the vibrations:

$$\begin{aligned} q_1^{(1)} &= A_1^{(1)} \sin\left(\omega_1 t + \beta_1\right) \\ q_2^{(1)} &= A_2^{(1)} \sin\left(\omega_1 t + \beta_1\right) \end{aligned} \quad \text{. . . . . . . . . . . .(7.7.)}$$

For $\omega_2$ there is the second main form of the vibrations:

$$\begin{aligned} q_1^{(2)} &= A_1^{(2)} \sin\left(\omega_2 t + \beta_2\right) \\ q_2^{(2)} &= A_2^{(2)} \sin\left(\omega_2 t + \beta_2\right) \end{aligned} \quad \text{. . . . . . . . . . (7.8.)}$$

$q_1^{(1)}, q_2^{(1)}, q_1^{(2)}$ and $q_2^{(2)}$ are the partial solution. The general solution is the sum of the partial solutions:

$$q_1 = q_1^{(1)} + q_1^{(2)} = A_1^{(1)} \sin\left(\omega_1 t + \beta_1\right) + A_1^{(2)} \sin\left(\omega_2 t + \beta_2\right)$$

$$q_2 = q_2^{(1)} + q_2^{(2)} = A_2^{(1)} \sin\left(\omega_1 t + \beta_1\right) + A_2^{(2)} \sin\left(\omega_2 t + \beta_2\right)$$

The constants $A_1^{(1)}$, $A_1^{(2)}$, $A_2^{(1)}$, $A_2^{(2)}$, $\beta_1$ and $\beta_2$ are unknown. It is necessary to find them from the initial conditions:

$q_1 = (q_1)_0$; $q_2 = (q_2)_0$; $\dot{q}_1 = (\dot{q}_1)_0$ and $\dot{q}_2 = (\dot{q}_2)_0$. So, there are six unknown constants and four equations (the initial conditions). It is possible to solve this problem with the introduction of the vibration's form coefficient $\mu$. It gives the interdependence between $A_2^{(1)}$ and $A_1^{(2)}$, i.e. $A_2^{(2)}$ and $A_1^{(2)}$.

$$\mu = \frac{q_2}{q_1} = \frac{A_2}{A_1} = -\frac{c_{11} - a_{11}\omega^2}{c_{12} - a_{12}\omega^2} \qquad \ldots\ldots (7.9.)$$

The coefficient $\mu$ depends upon the parameters of system (the coefficients of rigidity, inertia and circular frequency, as seen in equation (7.9.)).

The coefficient of vibration's form for the first main vibration:

$$\mu^{(1)} = \frac{A_2^{(1)}}{A_1^{(1)}} = \frac{q_2^{(1)}}{q_1^{(1)}} = -\frac{c_{11} - a_{11}\omega_1^2}{c_{12} - a_{12}\omega_1^2} \qquad \ldots . (7.10.)$$

The coefficient of vibration's form for the second main vibration:

$$\mu^{(2)} = \frac{A_2^{(2)}}{A_1^{(2)}} = \frac{q_2^{(2)}}{q_1^{(2)}} = -\frac{c_{11} - a_{11}\omega_2^2}{c_{12} - a_{12}\omega_2^2} \qquad \ldots . (7.11.)$$

Now, it is possible to write the general solution of the differential equation's system (7.2.) in form:

$$\begin{aligned} q_1 &= A_1^{(1)} \sin(\omega_1 t + \beta_1) + A_1^{(2)} \sin(\omega_2 t + \beta_2) \\ q_2 &= \mu^{(1)} A_1^{(1)} \sin(\omega_1 t + \beta_1) + \mu^{(2)} A_1^{(2)} \sin(\omega_2 t + \beta_2) \end{aligned} \qquad ..(7.12.)$$

Now the unknown constants are $A_1^{(1)}$, $A_1^{(2)}$, $\beta_1$ and $\beta_2$.

## 7.2. Problems

7.1. The homogeneous disc has mass *M* and radius *R*. It rolls on the horizontal ground without sliding and it is linked for the pendulum with the spring (Figure 7.1.). The spring has rigidity *c*. The pendulum is hung in point $O_1$ and its mass is neglected. There are two concentrated masses (*m* and 2*m*) on the pendulum. Estimate the system of differential equations, the circular frequencies of vibrations and the laws of vibration. It is given: $l$=2[m]; *R*=1[m]; *c*=1000[N/m]; *M*=30[kg]; *m*=10[kg], the initial conditions: $t_0$=0[s]; $(\varphi_1)_0 = 0[rad]$; $(\varphi_2)_0 = 0[rad]$; $(\dot{\varphi}_1)_0 = 1[rad/s]$; $(\dot{\varphi}_2)_0 = 0{,}8[rad/s]$.

Solution:

The kinetic and potential energy (Figure 7.2.):

$$E_k = E_{km} + E_{k2m} + E_{kM}$$

$$E_k = \frac{1}{2}m(l\dot{\varphi}_2)^2 + \frac{1}{2}2m(2l\dot{\varphi}_2)^2 + E_{KM}$$

$$E_{kM} = E_{kM}^{rot} + E_{kM}^{tr} = \frac{1}{2}J_0\dot{\varphi}_1^2 + \frac{1}{2}Mv_0^2$$

$$E_k = \left(\frac{1}{2}ml^2 + 4ml^2\right)\dot{\varphi}_2^2 + \frac{1}{2}\left(\frac{1}{2}MR^2\right)\dot{\varphi}_1^2 + \frac{1}{2}M(R\dot{\varphi}_1)^2$$

$$E_k = \frac{9}{2}ml^2\dot{\varphi}_2^2 + \frac{3}{4}MR^2\dot{\varphi}_1^2$$

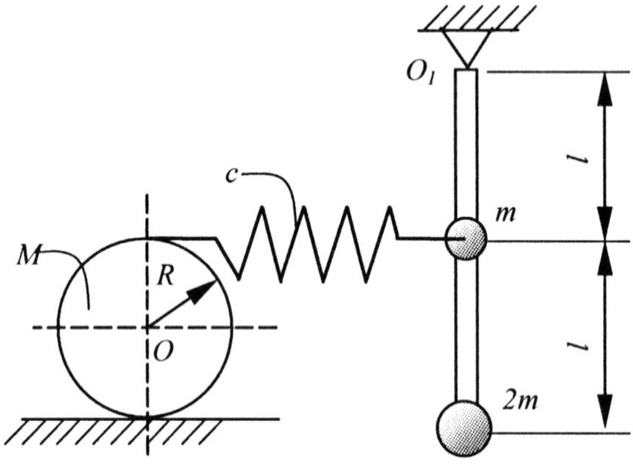


*Figure 7.1.*

$$E_p = E_{pm} + E_{p2m} + E_{pc}$$

$$E_{pm} = mg(l - l\cos\varphi_2);\ E_{p2m} = 2mg(2l - 2l\cos\varphi_2)$$

$$E_{pc} = \frac{1}{2}c(2R\varphi_1 - l\sin\varphi_2)^2$$

We will use LaGrange's equations and for small vibrations $\sin\varphi \approx \varphi$ and $\cos\varphi \approx 1 - \frac{\varphi^2}{2}$:

$$E_p \approx mgl\left(1-1+\frac{\varphi_2^2}{2}\right)+2mg\,2l\left(1-1+\frac{\varphi_2^2}{2}\right)+\frac{1}{2}c\left(2R\varphi_1-l\varphi_2\right)^2$$

$$E_p=\frac{5}{2}mgl\varphi_2^2+\frac{1}{2}c\left(2R\varphi_1-l\varphi_2\right)^2$$

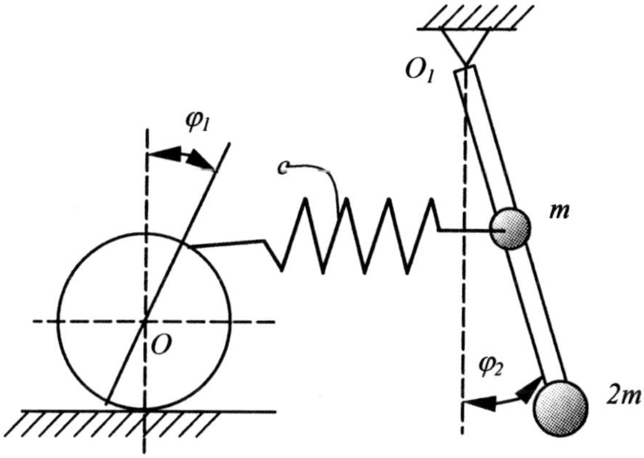


*Figure 7.2.*

LaGrange's equations have the form (7.1.):

$$\frac{d}{dt}\left(\frac{\partial E_k}{\partial \dot{q}_1}\right)-\frac{\partial E_k}{\partial q_1}+\frac{\partial E_p}{\partial q_1}=0$$

$$\frac{d}{dt}\left(\frac{\partial E_k}{\partial \dot{q}_2}\right)-\frac{\partial E_k}{\partial q_2}+\frac{\partial E_p}{\partial q_2}=0$$

$$2\frac{3}{4}MR^2\ddot{\varphi}_1+2\frac{1}{2}c\left(2R\varphi_1-l\varphi_2\right)2R=0$$

$$2\frac{9}{2}ml^2\ddot{\varphi}_2+2\frac{5}{2}mgl\varphi_2+2\frac{1}{2}c\left(2R\varphi_1-l\varphi_2\right)l=0$$

$$\frac{3}{2}MR^2\ddot{\varphi}_1+4cR^2\varphi_1-2Rcl\varphi_2=0$$

$$9ml^2\ddot{\varphi}_2+5mgl\varphi_2+2clR\varphi_1-cl^2\varphi_2=0 \qquad \ldots.(7.2.1.)$$

The partial solutions have the form (7.3.):

$$\varphi_1=A_1\cos(\omega t-\alpha);\quad \ddot{\varphi}_1=-\omega^2A_1\cos(\omega t-\alpha)$$

$$\varphi_2=A_2\cos(\omega t-\alpha);\quad \ddot{\varphi}_2=-\omega^2A_2\cos(\omega t-\alpha)$$

Put it in the system of differential equations (7.2.1.)

$$-\frac{3}{2}MR^2\omega^2A_1+4cR^2A_1-2RclA_2=0$$

$$-9ml^2\omega^2A_2+5mglA_2+c2lRA_1-cl^2A_2=0$$

$$\left(4cR^2-\frac{3}{2}MR^2\omega^2\right)A_1-2RclA_2=0 \qquad \ldots\ldots (7.2.1.1.)$$

$$2clRA_1+\left(5mgl-9ml^2\omega^2-cl^2\right)A_2=0$$

$$\Delta\left(\omega^2\right)=\begin{vmatrix}4cR^2-\frac{3}{2}MR^2\omega^2 & -2Rcl\\ 2clR & 5mgl-9ml^2\omega^2-cl^2\end{vmatrix}=0$$

The frequent equation:

$$\left(4cR^2-\frac{3}{2}MR^2\omega^2\right)\cdot\left(5mgl-9ml^2\omega^2-cl^2\right)+\left(2clR\right)^2=0$$

When we put the values in the frequent equation:

$$16200\cdot\omega^4-1304145\cdot\omega^2+3924000=0$$

It is biquadratic equation and its solutions are:

$$\omega_{1,2}^2=\frac{1304145\pm\sqrt{1304145^2-4\cdot16200\cdot3924000}}{2\cdot16200}$$

$\omega_1^2=3{,}13\left[s^{-2}\right]$ and $\omega_2^2=77{,}37\left[s^{-2}\right]$

$\omega_1=1{,}77\left[s^{-1}\right]$ and $\omega_2=8{,}8\left[s^{-1}\right]$

The laws of vibration have form (7.12.):

$$\varphi_1=A_1^{(1)}\sin\left(\omega_1t+\beta_1\right)+A_1^{(2)}\sin\left(\omega_2t+\beta_2\right) \qquad \ldots\ldots\ldots\ldots (I)$$

$$\varphi_2=\mu^{(1)}A_1^{(1)}\sin\left(\omega_1t+\beta_1\right)+\mu^{(2)}A_1^{(2)}\sin\left(\omega_2t+\beta_2\right) \qquad \ldots\ldots (II)$$

The unknown constants are $A_1^{(1)}$, $A_1^{(2)}$, $\beta_1$ and $\beta_2$.

$$\dot\varphi_1=A_1^{(1)}\omega_1\cos\left(\omega_1t+\beta_1\right)+A_1^{(2)}\omega_2\cos\left(\omega_2t+\beta_2\right) \qquad \ldots\ldots\ldots (III)$$

$$\dot\varphi_2=\mu^{(1)}A_1^{(1)}\omega_1\cos\left(\omega_1t+\beta_1\right)+\mu^{(2)}A_1^{(2)}\omega_2\cos\left(\omega_2t+\beta_2\right)\ldots (IV)$$

The vibration's form coefficient $\mu$:

$$\mu=\frac{A_2}{A_1}$$

From (7.2.1.1.):

$$\mu=\frac{A_2}{A_1}=\frac{4cR^2-\frac{3}{2}MR^2\omega^2}{2Rcl}$$

$$\mu^{(1)} = \frac{4cR^2 - \frac{3}{2}MR^2\omega_1^2}{2Rcl} = 0,96475$$

$$\mu^{(2)} = \frac{4cR^2 - \frac{3}{2}MR^2\omega_2^2}{2Rcl} = 0,1288$$

Put the initial conditions and $\mu^{(1)}$, $\mu^{(2)}$ in the equations (*I*), (*II*), (*III*) and (*IV*):

$$0 = A_1^{(1)} \sin\beta_1 + A_1^{(2)} \sin\beta_2$$

$$0 = \mu^{(1)} A_1^{(1)} \sin\beta_1 + \mu^{(2)} A_1^{(2)} \sin\beta_2$$

$$1 = A_1^{(1)} \cdot 1,77 \cdot \cos\beta_1 + A_1^{(2)} \cdot 8,8 \cdot \cos\beta_2$$

$$0,8 = 0,96475 \cdot A_1^{(1)} \cdot 1,77 \cdot \cos\beta_1 + 0,1288 \cdot A_1^{(2)} \cdot 8,8 \cdot \cos\beta_2$$

It is the system of four equations with four unknown constants. The solutions are:

$$A_1^{(1)} = 0,4536264604$$

$$\beta_1 = 0$$

$$A_1^{(2)} = 0,02239558695$$

$$\beta_2 = 0$$

The general solutions (laws of vibration):

$$\varphi_1 = 0,4536264604 \cdot \sin(1,77t) + 0,02239558695 \cdot \sin(8,8t)$$

$$\varphi_2 = 0,4376361276709 \cdot \sin(1,77t) + 0,00288455159916 \cdot \sin(8,8t)$$

7.2. There are two concentrated masses (*m* and 2*m*) on the lever (Figure 7.3.). the lever rotates around the axis in point *O* and its mass is neglected. The disc has mass *M* and it rotates around the axis in point $O_1$. There are three springs in vibration's system and they have rigidities $c_1$, $c_2$ and $c_3$. All resistances to motions are neglected. Estimate the system of differential equations using the law of change of angular momentum.

Solution:

It is necessary to make the imaginary cut of system in the spring with rigidity $c_2$ and to observe two systems with the forces (Figure 7.4.). The law of change of angular momentum for the system with the lever:

$$J_0\ddot{\varphi}_1 = \sum M_O^{Fi}$$

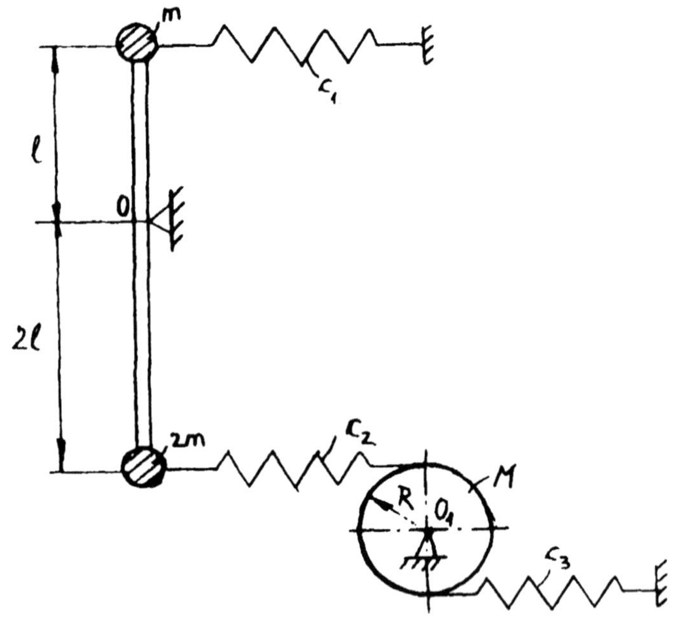


*Figure 7.3.*

In Figure 7.4:

$$J_0\ddot{\varphi}_1 = -F_{C1}l - F_{C2}2l + mgl\varphi_1 - 2mg2l\varphi_1$$

It is the first differential equation.

The law of change of angular momentum for the system with the disc:

$$J_{01}\ddot{\varphi}_2 = \sum M_{O1}^{Fi}$$

In Figure 7.4:

$$J_{O1}\ddot{\varphi}_2 = -F_{C2}R - F_{C3}R$$

It is the second differential equation.

$$F_{C1} = c_1 l\varphi_1; F_{C2} = c_2(2l\varphi_1 + R\varphi_2); F_{C3} = c_3 R\varphi_2$$

$$J_0 = ml^2 + 2m(2l)^2; J_{01} = \frac{1}{2}MR^2$$

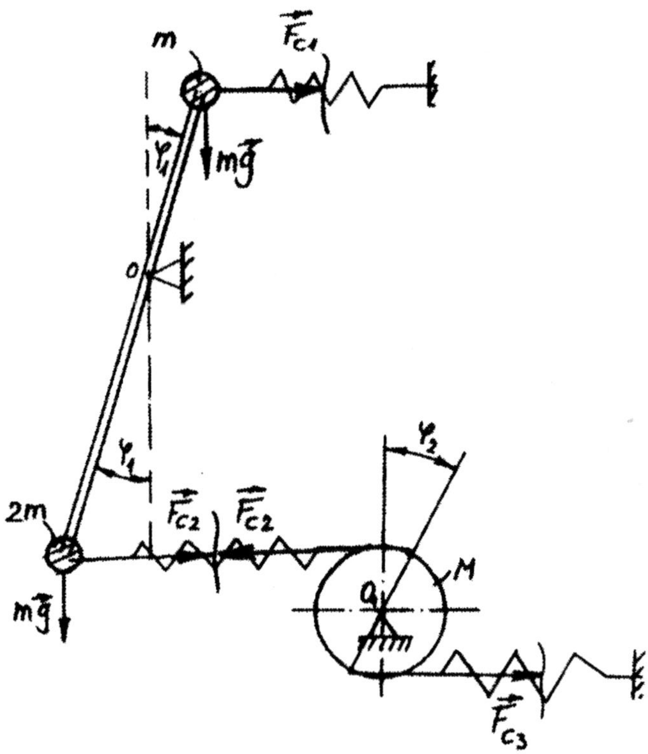


*Figure 7.4.*

$$\left[ml^2 + 2m(2l)^2\right] \cdot \ddot{\varphi}_1 = -c_1 l^2 \varphi_1 - c_2(2l\varphi_1 + R\varphi_2)2l - 3mgl\,\varphi_1$$

$$-\frac{1}{2}MR^2\ddot{\varphi}_2 = -c_2(2l\varphi_1 + R\varphi_2)R - c_3 R^2 \varphi_2$$

$$9ml^2\ddot{\varphi}_1 + c_1 l^2 \varphi_1 + c_2 4l^2 \varphi_1 + c_2 R 2l \varphi_2 + 3mgl\,\varphi_1 = 0$$

$$\frac{1}{2}MR^2\ddot{\varphi}_2 + c_2 2lR\varphi_1 + c_2 R^2 \varphi_2 + c_3 R^2 \varphi_2 = 0 \quad \text{.....(7.2.2.)}$$

7.3. Solve the previous problem using LaGrange's equations of second order.

The kinetic and potential energy (figure 7.4.):

$$E_k = E_{km} + E_{k2m} + E_{kM} = \frac{1}{2}m(l\dot{\varphi}_1)^2 + \frac{1}{2}2m(2l\dot{\varphi}_1)^2 + \frac{1}{2}J_{01}\dot{\varphi}_2 =$$

$$=\frac{9}{2}ml^2\dot{\varphi}_1+\frac{1}{4}MR^2\dot{\varphi}_2^2$$

$$E_p=E_{pc1}+E_{pc2}+E_{pc3}+E_{pm}+E_{p2m}=$$

$$=\frac{1}{2}c_1(l\varphi_1)^2+\frac{1}{2}c_2(R\varphi_2+2l\varphi_1)^2+\frac{1}{2}c_3(R\varphi_2)^2-mg(l-l\cos\varphi_1)+$$

$$+2mg(2l-2l\cos\varphi_1)$$

$$E_p\approx\frac{1}{2}c_1l^2\varphi_1^2+\frac{1}{2}c_2(R\varphi_2+2l\varphi_1)^2+\frac{1}{2}c_3R^2\varphi_2^2-mgl\frac{\varphi_1^2}{2}+4mgl\frac{\varphi_1^2}{2}$$

It is needed to put $E_k$ and $E_p$ in LaGrange's equations (7.1.):

$$\frac{d}{dt}\left(\frac{\partial E_k}{\partial\dot{q}_1}\right)-\frac{\partial E_k}{\partial q_1}+\frac{\partial E_p}{\partial q_1}=0$$

$$\frac{d}{dt}\left(\frac{\partial E_k}{\partial\dot{q}_2}\right)-\frac{\partial E_k}{\partial q_2}+\frac{\partial E_p}{\partial q_2}=0$$

---

and we will get the system of differential equations:

$$9ml^2\ddot{\varphi}_1+c_1l^2\varphi_1+c_2(R\varphi_2+2l\varphi_1)2l+3mgl\varphi_1=0$$

$$\frac{1}{2}MR^2\ddot{\varphi}_2+c_2(R\varphi_2+2l\varphi_1)R+c_3R^2\varphi_2=0$$

---

$$9ml^2\ddot{\varphi}_1+c_1l^2\varphi_1+c_2R\varphi_2 2l+c_2 4l^2\varphi_1+3mgl\varphi_1=0$$

$$\frac{1}{2}MR^2\ddot{\varphi}_2+c_2 2lR\varphi_1+c_2R^2\varphi_2+c_3R^2\varphi_2=0$$

---

The last system of two differential equations is equal to (7.2.2.).

7.4. The sheet metal has mass $m_2$ and it does the motion of translation between four barrels. The barrels are the same and they have mass *m* and radius *r*. The end of the sheet is linked to an immovable wall with the spring. The other end is linked to the disc with a spring. The springs have rigidity *c*. The disc has mass $m_1$ and radius *R* (Figure 7.5.). Find the frequent equation of the system's vibrations, if $m_1$=$m_2$=2m.

Solution:

The kinetic and potential energy (Figure 7.6.):

$$E_k = E_{km1} + E_{km2} + 4E_{km}$$

$$E_{km1} = \frac{1}{2}J_0\dot{\varphi}_1^2 = \frac{1}{2}\left(\frac{1}{2}m_1R^2\right)\dot{\varphi}_1^2; E_{km2} = \frac{1}{2}m_2\dot{x}^2$$

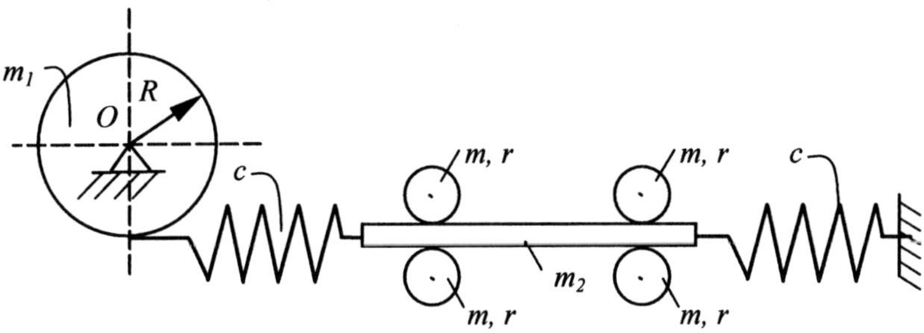


Figure 7.5.

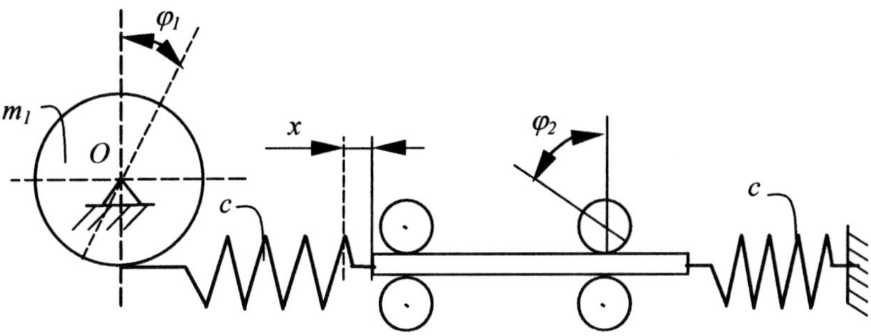


Figure 7.6.

$$E_{km} = \frac{1}{2}J_{O1}\dot{\varphi}_2^2 = \frac{1}{2}\left(\frac{1}{2}mr^2\right)\dot{\varphi}_2^2$$

$$E_k = \frac{1}{4}m_1R^2\dot{\varphi}_1^2 + \frac{1}{2}m_2\dot{x}^2 + 4\frac{1}{4}mr^2\dot{\varphi}_2^2$$

$$E_k = \frac{m_1R^2}{4}\dot{\varphi}_1^2 + \frac{m_2}{2}\dot{x}^2 + mr^2\dot{\varphi}_2^2$$

The interdependence of coordinates $\varphi_2$ and $x$:

$$r\varphi_2 = x \quad \Rightarrow \quad r\cdot\frac{d\varphi_2}{dt} = \frac{dx}{dt} \quad \Leftrightarrow \quad r\dot{\varphi}_2 = \dot{x} \quad \Rightarrow \quad \dot{\varphi}_2 = \frac{\dot{x}}{r}$$

$$E_k = \frac{m_1R^2}{4}\dot{\varphi}_1^2 + \left(\frac{m_2}{2} + m\right)\dot{x}^2$$

The potential energy is composed of potential energy of the left ($E_{pcL}$) and right ($E_{pcD}$) spring (Figure 7.6.):

$$E_p = E_{pcL} + E_{pcD} = \frac{1}{2}c(R\varphi_1 + x)^2 + \frac{1}{2}cx^2$$

LaGrange's equations:

$$\frac{d}{dt}\left(\frac{\partial E_k}{\partial \dot{\varphi}_1}\right) - \frac{\partial E_k}{\partial \varphi_1} + \frac{\partial E_p}{\partial \varphi_1} = 0$$

$$\frac{d}{dt}\left(\frac{\partial E_k}{\partial \dot{x}}\right) - \frac{\partial E_k}{\partial x} + \frac{\partial E_p}{\partial x} = 0$$

$$\frac{m_1 R^2}{2}\ddot{\varphi}_1 - 0 + c(R\varphi_1 + x)R = 0$$

$$2\left(\frac{m_2}{2} + m\right)\ddot{x} - 0 + c(R\varphi_1 + x) + cx = 0$$

$$\frac{m_1 R^2}{2}\ddot{\varphi}_1 + cR^2\varphi_1 + cRx = 0 \qquad \ldots\ldots (7.2.3.)$$

$$(m_2 + 2m)\ddot{x} + cR\varphi_1 + 2cx = 0$$

The partial solutions have form (7.3.):

$\varphi_1 = A_1\cos(\omega t + \beta); \ddot{\varphi}_1 = -A_1\omega^2\cos(\omega t + \beta)$

$x = A_2\cos(\omega t + \beta); \ddot{x} = -A_2\omega^2\cos(\omega t + \beta)$

Put it in the system (7.2.3.) and abbreviate the equations with $\cos(\omega t + \beta)$:

$$-\frac{m_1 R^2}{2}A_1\omega^2 + cR^2 A_1 + cRA_2 = 0$$

$$-(m_2 + 2m)A_2\omega^2 + cRA_1 + 2cA_2 = 0$$

$$\left(cR - \frac{m_1 R\omega^2}{2}\right)A_1 + cA_2 = 0$$

$$cRA_1 + \left(2c - m_2\omega^2 - 2m\omega^2\right)A_2 = 0$$

$$\Delta(\omega^2) = \begin{vmatrix} cR - \frac{m_1 R}{2}\omega^2 & c \\ cR & 2c - m_2\omega^2 - 2m\omega^2 \end{vmatrix} = 0$$

If we develop the determinant, we will get the frequent equation:

$$\left(cR - \frac{m_1 R}{2}\omega^2\right)\cdot\left(2c - m_2\omega^2 - 2m\omega^2\right) - c^2 R = 0$$

set it in order and put $m_1 = m_2 = 2m$:

$$c^2 R + \frac{mR}{2}(2m + 2m)\omega^4 - cR\omega^2(2m + 2m) - mRc\,\omega^2 = 0$$

$$2m^2 R\omega^4 - 5cRm\omega^2 + c^2 R = 0.$$

7.5. A disc (mass $M=2m$ and radius $R$) rotates round the axis in point $O$. A bar (mass $m$ and length $l=2R$) is linked for the disc with joint in point $O_1$. Estimate the system of differential equations. All frictions are neglected.

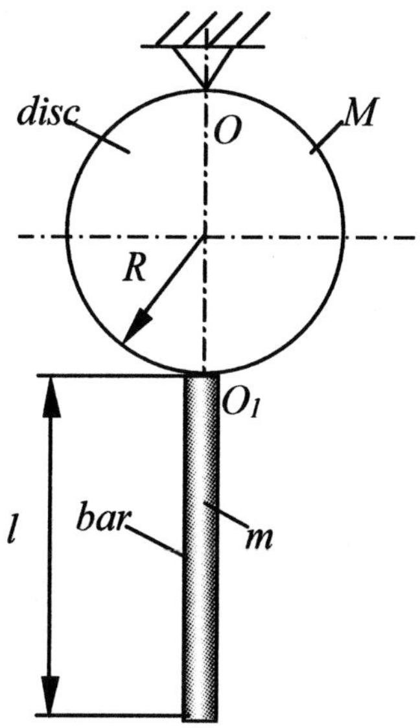


*Figure 7.7.*

Final solution:

$$7R\ddot{\varphi}_1 + 4g\varphi_1 + 2R\ddot{\varphi}_2 = 0$$

$$2R\ddot{\varphi}_1 + g\varphi_2 + \frac{4}{3}\ddot{\varphi}_2 = 0$$

---

7.6. A bar AB has mass 2m and it moves without sliding on two discs (Figure 7.8.). The discs have mass $m$, radius $r$ and they rotate round the individual axes. Unextensible rope with concentrated mass *M* is hooked up in point *C* of the bar. Find the frequent equation of the system's vibrations.

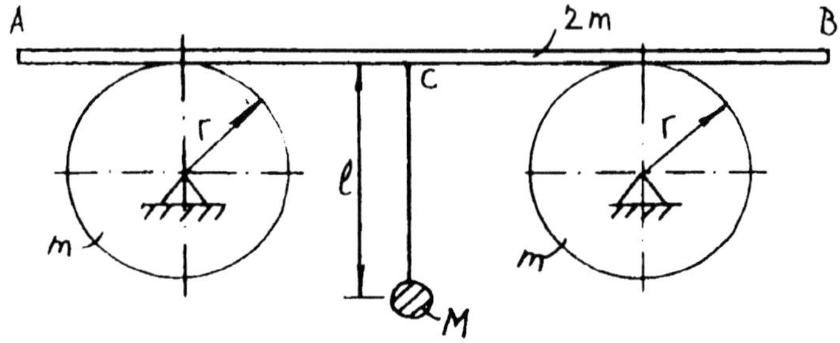


*Figure 7.8.*

Final solution:

$$3mMr^2l^2\Omega^4 - \left(3mMr^2l \cdot g + r^2gM^2l\right)\Omega^2 = 0$$

(Hint: See problem 9.3.)

# CHAPTER EIGHT

## TRANSVERSE VIBRATIONS OF CONCENTRATED MASSES ON THE LIGHT BEAM WITH TWO OR MORE DEGREES OF VIBRATION'S FREEDOM

### 8.1. Short Theory Review

It is necessary to take into consideration the next presumptions on the occasion of solving the problems:
- longitudinal displacements of the masses are neglected and it is needed to take into consideration just transversal once,
- mass of the beam is neglected,
- the system vibrates about the position of static equilibrium,
- the system vibrates in the form of beam's static elastic line.

#### 8.1.1. Statically Determinate Beams

We can use the influence coefficients (Maxwell - Mohr coefficients) $\alpha_{ij}$ for getting the differential equations of vibrations. The influence coefficient $\alpha_{ij}$ is the bilge of beam in point "*i*" when unit force (*F*=1) acts upon the beam in point "*j*".
If there are two concentrated masses on the beam in points ① and ② (Figure 8.1.), then the bilge in point ① depends upon the forces in points ① and, also, ②.

$$y_1 = \alpha_{11}F_1 + \alpha_{12}F_2 \qquad \text{(8.1.1.)}$$

where
- $y_1$ is the bilge in point ①;
- $\alpha_{11}$ is the influence (Maxwell - Mohr) coefficient, which is the bilge in point ①, when the unit force acts upon point ①;
- $\alpha_{12}$ is the influence coefficient, which is the bilge in point ①, when the unit force acts upon point ②;
- $F_1$ is the force in point ① including the inertial force;
- $F_2$ is the force in point ② including the inertial force.
Analogously, the bilge $y_2$ in point ②:

$$y_2 = \alpha_{21}F_1 + \alpha_{22}F_2 \qquad \text{(8.1.2.)}$$

It is needed to say that $\alpha_{ij} = \alpha_{ji}$ and it is possible to choose the influence coefficients in tables (the tables 1 to 5) of the appendix. The equations (8.1.1.) and (8.1.2.) are the system of two differential equations. It is possible to solve the system the way it is described in chapter six.
The number of concentrated masses on the beam is the number of freedom's degrees, and of differential equations in the system.

### 8.1.2. Statically Indeterminate Beams

The influence coefficients for statically indeterminate beams have a mark with asterisk ($\alpha^*_{ij}$). We can choose $\alpha^*_{ij}$ in the tables of appendix. Also: $\alpha^*_{ij}=\alpha^*_{ji}$. It is possible to find the influence coefficients from additional condition about equality of the bilges, if the beam is composed of more beams.

## 8.2. Problems

8.1. There are two concentrated masses on the cantilever beam and mass *m* is linked to the ground with the spring (Figure 8.1.). The spring has rigidity *c*. It is needed to find the frequencies of vibrations $\omega_1$ and $\omega_2$, if it is known: the beam is I-iron I10 ($E=2\cdot10^{11}$[N/m$^2$]; $I=171\cdot10^{-8}$[m$^4$]); $l=10[m]$; $c=1000$ [N/m]; $m=100$[kg].

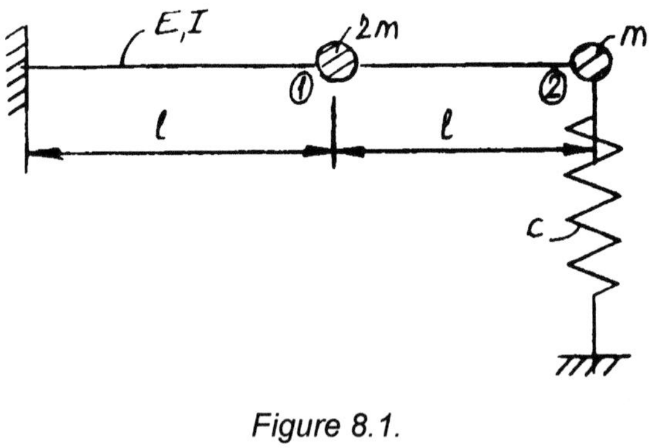


*Figure 8.1.*

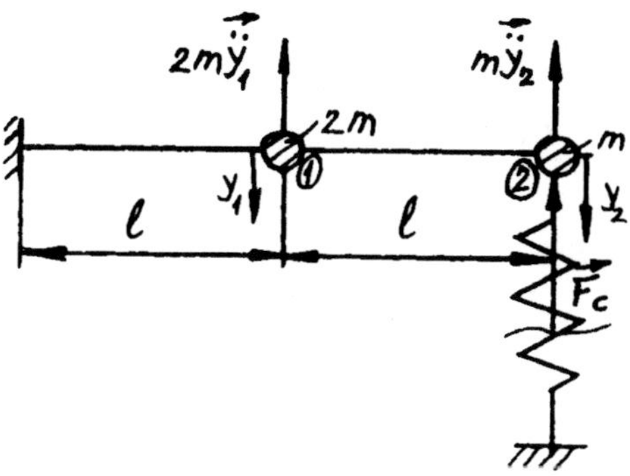


*Figure 8.2.*

Solution:

The bilge in points ① and ②:

$$y_1 = \alpha_{11}F_1 + \alpha_{12}F_2 = \alpha_{11}\cdot(-2m\ddot{y}_1) + \alpha_{12}\cdot(-m\ddot{y}_2 - F_c)$$
$$y_2 = \alpha_{21}F_1 + \alpha_{22}F_2 = \alpha_{21}\cdot(-2m\ddot{y}_1) + \alpha_{22}\cdot(-m\ddot{y}_2 - F_c)$$

$F_1$ and $F_2$ are the forces in points ① and ②. There is just the inertial force $2m\ddot{y}_1$ in point ①, and the inertial $m\ddot{y}_2$ and the force of spring $F_c$ in point ②.

The force of spring: $F_c = cy_2$

$$y_1 = -2m\alpha_{11}\ddot{y}_1 - \alpha_{12}m\ddot{y}_2 - \alpha_{12}cy_2$$
$$y_2 = -2m\alpha_{21}\ddot{y}_1 - \alpha_{22}m\ddot{y}_2 - \alpha_{22}cy_2$$

The system of differential equations:

$$\begin{aligned}-2m\alpha_{11}\ddot{y}_1 - m\alpha_{12}\ddot{y}_2 - y_1 - c\alpha_{12}y_2 = 0\\ -2m\alpha_{21}\ddot{y}_1 - m\alpha_{22}\ddot{y}_2 - y_2 - c\alpha_{22}y_2 = 0\end{aligned} \quad \ldots (8.2.1.)$$

The partial solutions have form:

$$y_1 = A_1\cos(\omega t - \alpha); \quad \ddot{y}_1 = -A_1\omega^2\cos(\omega t - \alpha)$$
$$y_2 = A_2\cos(\omega t - \alpha); \quad \ddot{y}_2 = -A_2\omega^2\cos(\omega t - \alpha)$$

Put it in the system (8.2.1.):

$$2m\alpha_{11}A_1\omega^2 + m\alpha_{12}A_2\omega^2 - A_1 - c\alpha_{12}A_2 = 0$$
$$2m\alpha_{21}A_1\omega^2 + m\alpha_{22}A_2\omega^2 - A_2 + c\alpha_{22}A_2 = 0$$

$$\left(2m\alpha_{11}\omega^2 - 1\right)A_1 + \left(m\alpha_{12}\omega^2 - c\alpha_{12}\right)A_2 = 0$$
$$2m\alpha_{21}\omega^2 A_1 + \left(m\alpha_{22}\omega^2 - 1 - c\alpha_{22}\right)A_2 = 0$$

$$\Delta\left(\omega^2\right) = \begin{vmatrix} 2m\alpha_{11}\omega^2 - 1 & m\alpha_{12}\omega^2 - c\alpha_{12} \\ 2m\alpha_{21}\omega^2 & m\alpha_{22}\omega^2 - 1 - c\alpha_{22} \end{vmatrix} = 0$$

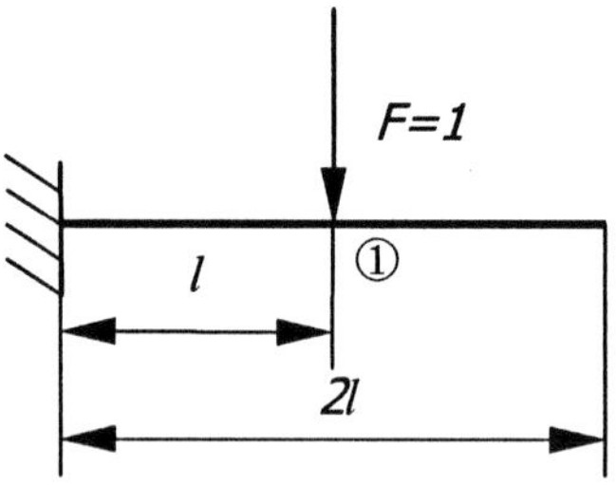


Figure 8.3.

The frequent equation:
$$\left(2m\alpha_{11}\omega^2 - 1\right)\left(m\alpha_{22}\omega^2 - 1 - c\alpha_{22}\right) - \left(m\alpha_{12}\omega^2 - c\alpha_{12}\right)\left(2m\alpha_{21}\omega^2\right) = 0$$
It is necessary to calculate influence coefficients $\alpha_{11}$, $\alpha_{12}=\alpha_{21}$ and $\alpha_{22}$. We can use the tables of appendix. Table 2. is for the cantilever beam. The unit force ($F$=1) acts upon the beam in point ① (Figure 8.3.) and we need the bilge in point ①. That case is in 3rd row of Table 2, so:

$$\alpha_{11} = \frac{q^3}{3EI}$$

In Figure 8.3: $q = l$ and: $\alpha_{11} = \dfrac{l^3}{3EI}$

$$\alpha_{11} = \frac{10^3\left[m^3\right]}{3\cdot 2\cdot 10^{11}\left[N/m^2\right]171\cdot 10^{-8}\left[m^4\right]} = 0{,}975\cdot 10^{-3}\left[\frac{m}{N}\right]$$

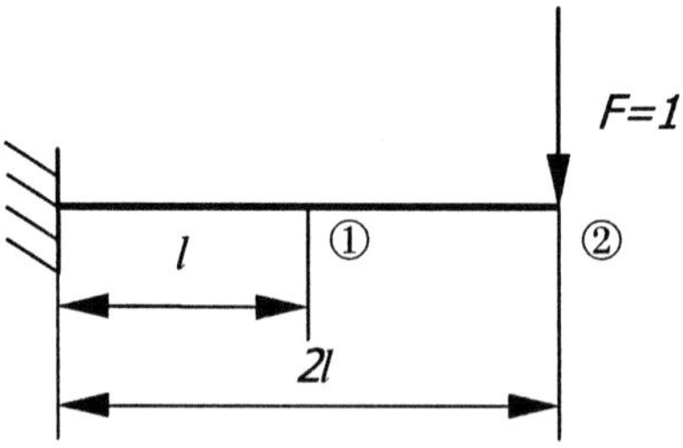


Figure 8.4.

For giving $\alpha_{12}$ we use Figure 8.4. (it is needed to estimate the bilge in point ① when the unit force acts upon point ②) and Table 2. (first row):

$$\alpha_{12} = \frac{(2l)^3}{6EI}\left[3\frac{2l}{2l}\left(\frac{l}{2l}\right)^2 - \left(\frac{l}{2l}\right)^3\right]$$

Put the values for $l$, $E$ and $I$:

$$\alpha_{12} = 6{,}092 \cdot 10^{-6}\left[\frac{m}{N}\right]$$

The coefficient $\alpha_{22}$ we got from Table 2. - 3rd row and Figure 8.5:

$$\alpha_{22} = \frac{q^3}{3EI},$$

but now $q = 2l$ in Figure 8.5.

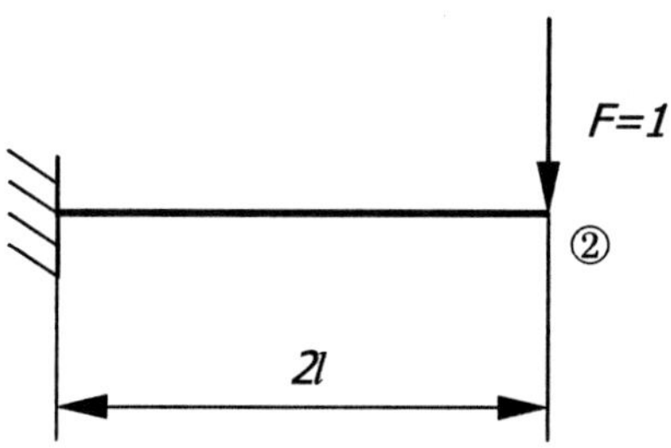


*Figure 8.5.*

$$\alpha_{22} = \frac{(2l)^3}{3EI}$$

$$\alpha_{22} = 7{,}797 \cdot 10^{-3}\left[\frac{m}{N}\right]$$

Put the values of influence coefficients and $m$, $c$ in the frequent equation:

$$0{,}152 \cdot \omega^4 - 2{,}495 \cdot \omega^2 + 8{,}797 = 0$$

The solutions of the last equation are:

$\omega_1^2 = 5{,}130\left[s^{-2}\right]$ and $\omega_2^2 = 11{,}284\left[s^{-2}\right]$

The frequencies are:

$\omega_1 = 2{,}265\left[s^{-1}\right]$ and $\omega_2 = 3{,}359\left[s^{-1}\right]$.

8.2. Estimate the frequent equation and the influence coefficients for vibration's system in Figure 8.6. It is known: $l$, $m$, $c$, $E$, and $I$.

Solution:
The system has three concentrated masses and three degrees of vibration's freedom. The first and second differential equations are:

$$y_1 = \alpha_{11} F_1 + \alpha_{12} F_2$$

$$y_1 = \alpha_{11} F_1 + \alpha_{12} F_2$$

$$y_1 = \alpha_{11}\left(-m\ddot{y}_1 + F_c\right) + \alpha_{12}\left(-m\ddot{y}_2\right) \quad ..(8.2.2.)$$

$$y_2 = \alpha_{21}\left(-m\ddot{y}_1 + F_c\right) + \alpha_{22}\left(-m\ddot{y}_2\right) \quad ..(8.2.3.)$$

It is necessary to imaginably cut the spring and observe the lower mass (Figure 8.7.). Newton's second law for the lower mass $m$:

$$m\ddot{y}_3 = -F_c$$

where $y_3$ is a feed of the lower mass (Figure 8.7.). The force of spring

$$F_c = c \cdot (y_3 - y_1)$$

Put it in Newton's second law:

$$m\ddot{y}_3 + c(y_3 - y_1) = 0 \quad \ldots\ldots\ldots\ldots\ldots (8.2.4.)$$

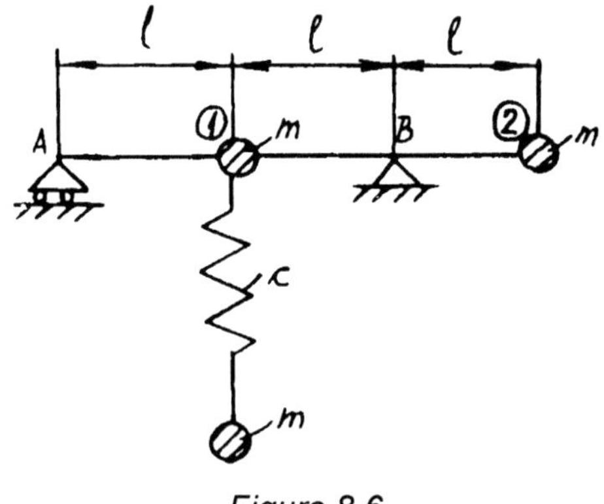


*Figure 8.6.*

The equations (8.2.2.), (8.2.3.) and (8.2.4.) are the system of three differential equations:

$$y_1 = \alpha_{11}\left(-m\ddot{y}_1 + c(y_3 - y_1)\right) + \alpha_{12}\left(-m\ddot{y}_2\right)$$
$$y_2 = \alpha_{21}\left(-m\ddot{y}_1 + c(y_3 - y_1)\right) + \alpha_{22}\left(-m\ddot{y}_2\right)$$
$$m\ddot{y}_3 + c(y_3 - y_1) = 0$$

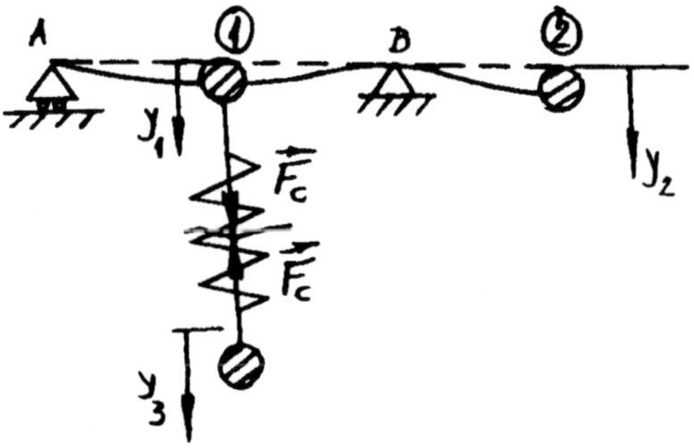


*Figure 8.7.*

The partial solutions are:

$y_i = A_i \cos(\omega t - \alpha)$,

$\ddot{y}_i = -A_i\omega^2 \cos(\omega t - \alpha)$,

where $i$ has values 1, 2 and 3.

Put the partial solutions $y_i$ and their derivatives $\ddot{y}_i$ in the system of differential equations and abbreviate all equations with cos($\omega t$-$\alpha$):

$$A_1 = \alpha_{11}\left[c(A_3 - A_1) + mA_1\omega^2\right] + \alpha_{12}mA_2\omega^2$$
$$A_2 = \alpha_{21}\left[c(A_3 - A_1) + mA_1\omega^2\right] + \alpha_{22}mA_2\omega^2$$
$$-mA_3\omega^2 + c(A_3 - A_1) = 0$$

If we set the system in order, we will get:

$$\left(1 + \alpha_{11}c - \alpha_{11}m\omega^2\right)A_1 - \alpha_{12}m\omega^2 A_2 - \alpha_{11}cA_3 = 0$$
$$\left(\alpha_{21}c - \alpha_{21}m\omega^2\right)A_1 + \left(1 - \alpha_{22}m\omega^2\right)A_2 - \alpha_{21}cA_3 = 0$$
$$-cA_1 + 0 \cdot A_2 + \left(c - m\omega^2\right)A_3 = 0$$

$$\Delta(\omega^2) = \begin{vmatrix} 1 + \alpha_{11}c - \alpha_{11}m\omega^2 & -\alpha_{12}m\omega^2 & -\alpha_{11}c \\ \alpha_{21}c - \alpha_{21}m\omega^2 & 1 - \alpha_{22}m\omega^2 & -\alpha_{21}c \\ -c & 0 & c - m\omega^2 \end{vmatrix} = 0$$

When we solve the determinant:

$$\left(1+\alpha_{11}c-\alpha_{11}m\omega^2\right)\left(c-m\omega^2-\alpha_{22}cm\omega^2+\alpha_{22}m^2\omega^4\right)+$$
$$+\alpha_{12}m\omega^2\left(\alpha_{12}c^2-\alpha_{21}cm\omega^2-\alpha_{21}cm\omega^2+\alpha_{21}m^2\omega^4-\alpha_{21}c^2\right)-$$
$$-\alpha_{11}c\left(c-\alpha_{22}cm\omega^2\right)=0$$

It is the frequent equation.

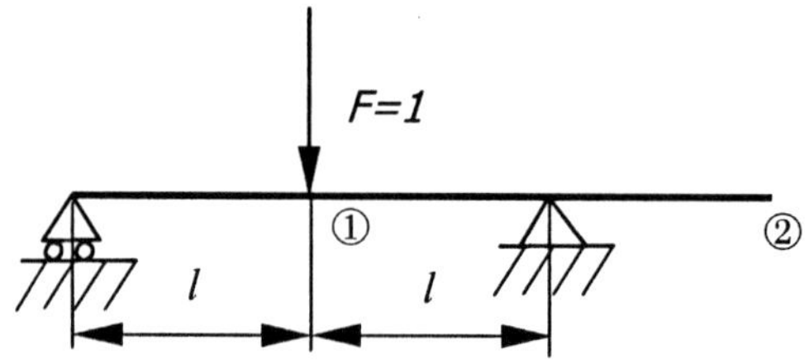


*Figure 8.8.*

It is needed to estimate the influence coefficients: $\alpha_{11}$, $\alpha_{12}=\alpha_{21}$ and $\alpha_{22}$. There is the beam with the overhang in the problem, so we will use Tables 1. and 3. of the appendix. For the coefficient $\alpha_{11}$ we will use Figure 8.8. and Table 1. (3rd row):

$$\alpha_{11}=\frac{(2l)^3}{3EI}\left(\frac{l}{2l}\right)^2\left(\frac{l}{2l}\right)^2=\frac{1}{6}\frac{l^3}{EI}$$

For $\alpha_{12}$ we will use Figure 8.9. and Table 3. (1st row):

$$\alpha_{12}=-\frac{(2l)^3}{6EI}\cdot\frac{l}{2l}\left[\frac{l}{2l}-\left(\frac{l}{2l}\right)^3\right],\text{ i.e. }\alpha_{12}=-\frac{1}{4}\frac{l^3}{EI}$$

For $\alpha_{22}$ we will use Figure 8.10. and Table 3. (3rd row):

$$\alpha_{22}=\frac{l^2}{3EI}(2l+l)=\frac{l^3}{EI}$$

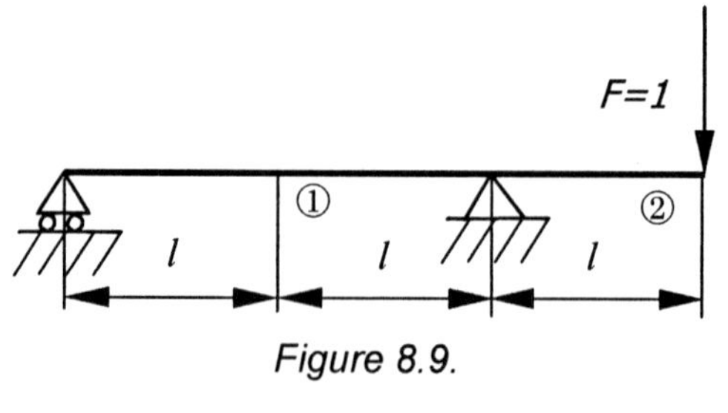


*Figure 8.9.*

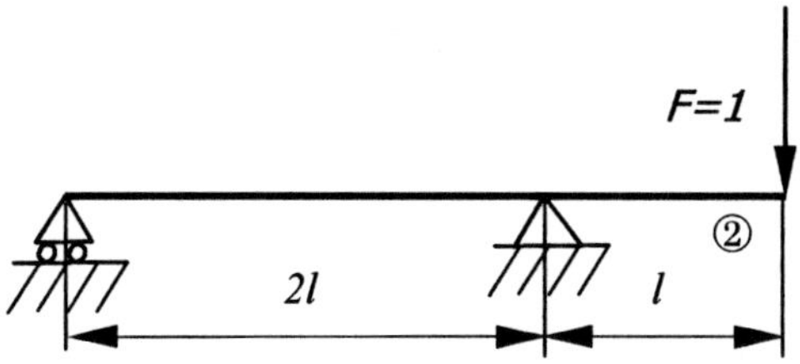


Figure 8.10.

8.3. Estimate the frequent equation of system in Figure 8.11, and influence coefficients. It is given: $l = 1[m]$; $E = 2 \cdot 10^{11}\left[\frac{N}{m^2}\right]$; the cross section of beam is the angle iron L70x70x9; $I = 52{,}6 \cdot 10^{-8}\left[m^4\right]$ (German standard DIN 1028).

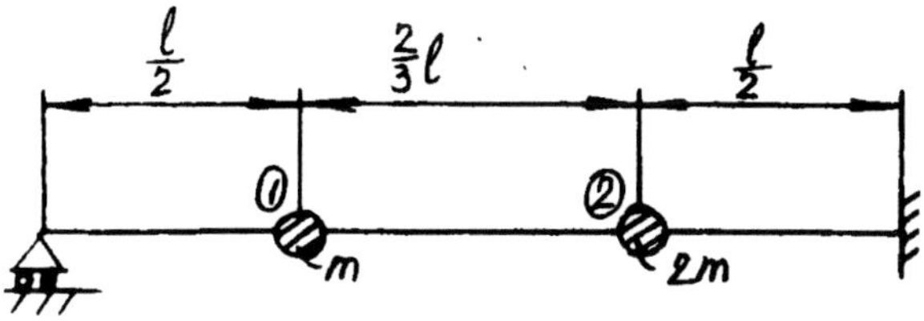

Figure 8.11.

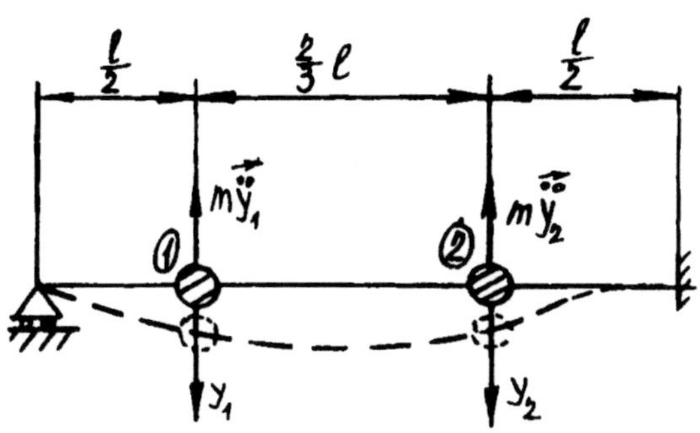

Figure 8.12.

The bilge in points ① and ②:

$$y_1 = \alpha_{11}^{*}(-m\ddot{y}_1) + \alpha_{12}^{*}(-2m\ddot{y}_2)$$

$$y_2 = \alpha_{21}^{*}(-m\ddot{y}_1) + \alpha_{22}^{*}(-2m\ddot{y}_2)$$

The partial solutions are:

$$y_1 = A_1 \cos(\omega t - \alpha);\ y_2 = A_2 \cos(\omega t - \alpha)$$

$\ddot{y}_1 = -A_1\omega^2\cos(\omega t - \alpha);\ddot{y}_2 = -\omega^2 A_2\cos(\omega t - \alpha)$

Put them in the system of differential equations and abbreviate the equations with cos($\omega t$ - $\alpha$):

$$\left(1 - m\alpha_{11}^{*}\omega^2\right)A_1 - 2m\alpha_{12}^{*}\omega^2 A_2 = 0$$

$$-m\alpha_{21}^{*}\omega^2 A_1 + \left(1 - 2m\alpha_{22}^{*}\omega^2\right)A_2 = 0$$

$$\Delta\left(\omega^2\right) = \begin{vmatrix} 1 - m\omega^2\alpha_{11}^{*} & -2m\alpha_{12}^{*}\omega^2 \\ -m\omega^2\alpha_{21}^{*} & 1 - 2m\omega^2\alpha_{22}^{*} \end{vmatrix} = 0$$

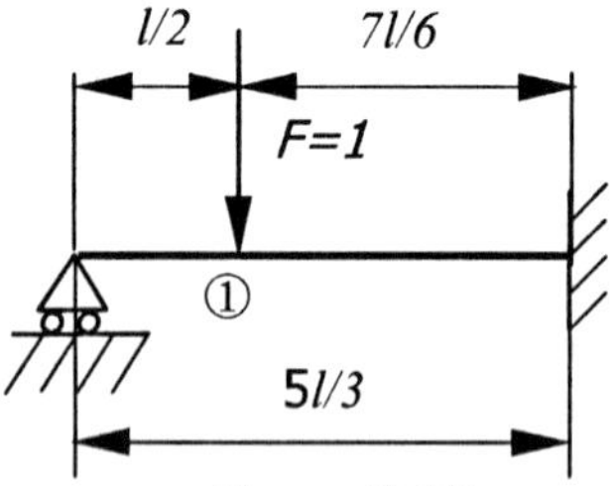


*Figure 8.13.*

We will get the frequent equation with solving the determinant:

$$\left(1 - m\omega^2\alpha_{11}^{*}\right)\left(1 - 2m\omega^2\alpha_{22}^{*}\right) - m\omega^2\alpha_{21}^{*}2m\alpha_{12}^{*}\omega^2 = 0$$

$$1 - 2m\omega^2\alpha_{22}^{*} - m\omega^2\alpha_{11}^{*} + 2m^2\omega^4\alpha_{11}^{*}\alpha_{22}^{*} - 2m^2\omega^4\alpha_{21}^{*\,2} = 0$$

It is needed to estimate $\alpha_{11}^{*}$ ; $\alpha_{12}^{*} = \alpha_{21}^{*}$ ; $\alpha_{22}^{*}$. The influence coefficient $\alpha_{11}^{*}$ is found using Figure 8.13. and Table 4 (3rd row)

$$\alpha_{11}^{*} = \frac{\left(\frac{5l}{3}\right)^3}{12EI}\left(\frac{\frac{7l}{6}}{\frac{5l}{3}}\right)^3\left(\frac{\frac{l}{2}}{\frac{5l}{3}}\right)^2\left(3 + \frac{\frac{l}{2}}{\frac{5l}{3}}\right)$$

Put the values for $l$, $E$ and $I$: $\alpha_{11}^{*} = 0{,}374\cdot 10^{-6}\left[\frac{m}{N}\right]$

$\alpha_{12}^{*} = \alpha_{21}^{*}$ is found using Figure 8.14. and Table 4. (2nd row):

$$\alpha_{12}^{*} = \frac{\left(\frac{5l}{3}\right)^3}{12\,EI}\left(\frac{\frac{l}{2}}{\frac{5l}{3}}\right)^2\left(1-\frac{\frac{3l}{6}+\frac{4l}{6}}{\frac{5l}{3}}\right)\cdot$$

$$\cdot\left[3\frac{\frac{7l}{6}}{\frac{5l}{3}}-\left(2+\frac{\frac{7l}{6}}{\frac{5l}{3}}\right)\cdot\left(1-\frac{\frac{l}{2}+\frac{2l}{3}}{\frac{5l}{3}}\right)^2\right]$$

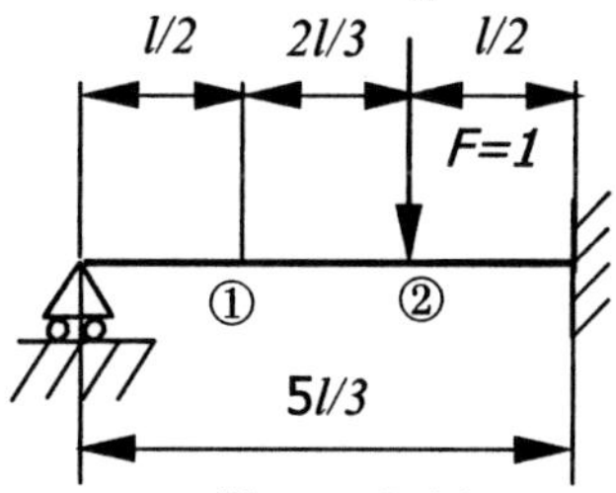


Figure 8.14.

$$\alpha_{12}^{*} = 1{,}0386\cdot 10^{-6}\left[\frac{m}{N}\right]$$

We can estimate $\alpha_{22}^{*}$ using Figure 8.15. and Table 4. (3[rd] row)

$$\alpha_{22}^{*} = \frac{\left(\frac{5l}{3}\right)^3}{12EI}\left(\frac{\frac{l}{2}}{\frac{5l}{3}}\right)^3\left(\frac{\frac{7l}{6}}{\frac{5l}{3}}\right)^2\left(3+\frac{\frac{7l}{6}}{\frac{5l}{3}}\right)$$

$$\alpha_{22}^{*} = 0{,}1795\cdot 10^{-6}\left[\frac{m}{N}\right]$$

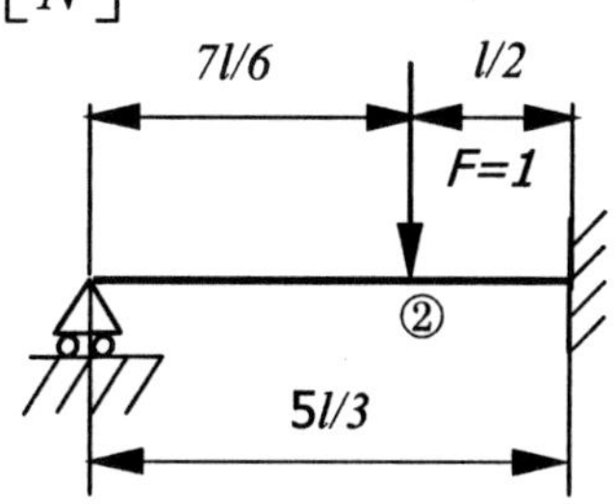


Figure 8.15.

8.4. Estimate the frequent equation of transverse vibrations for system in Figure 8.16. and the influence coefficients. The beams *AB* and *CD* are perpendicular and they are in the horizontal plane. They have fixed joint in point *C*.

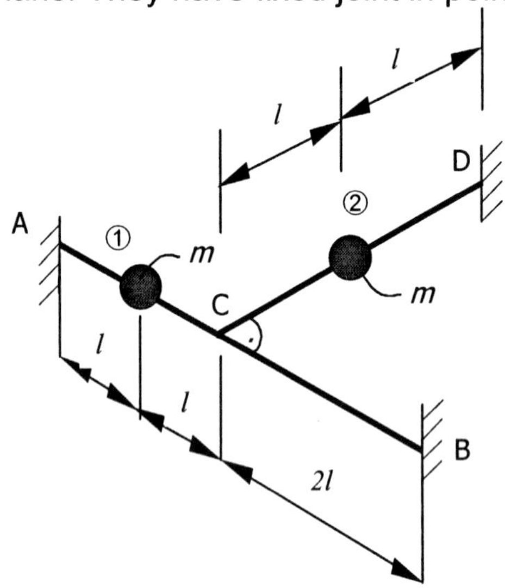


*Figure 8.16.*

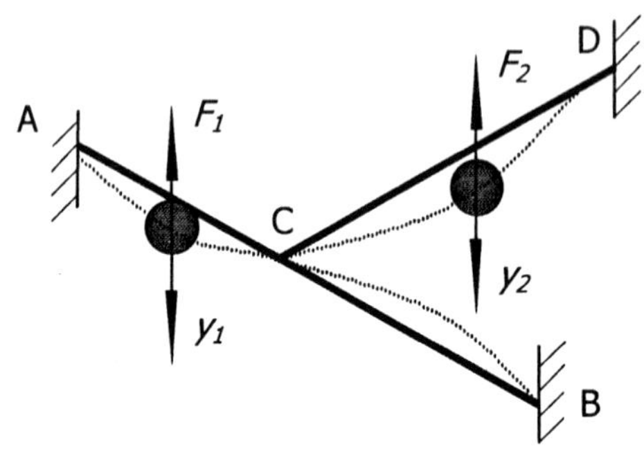


*Figure 8.17.*

Solution:
The system is statically indeterminate.
$y_1$ is the bilge in point ① and $y_2$ in point ② (Figure 8.16.):

$$y_1 = \alpha_{11}^* \cdot F_1 + \alpha_{12}^* \cdot F_2$$

$$y_2 = \alpha_{21}^* \cdot F_1 + \alpha_{22}^* \cdot F_2$$

The inertial forces: $F_1 = -m\ddot{y}_1$ and $F_2 = -m\ddot{y}_2$

$$y_1 = \alpha_{11}^* \cdot (-m\ddot{y}_1) + \alpha_{12}^* \cdot (-m\ddot{y}_2)$$
$$y_2 = \alpha_{21}^* \cdot (-m\ddot{y}_1) + \alpha_{22}^* \cdot (-m\ddot{y}_2)$$

$$y_1 + \alpha_{11}^* \cdot m\ddot{y}_1 + \alpha_{12}^* \cdot m\ddot{y}_2 = 0$$
$$y_2 + \alpha_{21}^* \cdot m\ddot{y}_1 + \alpha_{22}^* \cdot m\ddot{y}_2 = 0$$

The partial solutions are:

$$y_1 = A_1 \cos(\omega t - \alpha)$$
$$y_2 = A_2 \cos(\omega t - \alpha)$$
$$\ddot{y}_1 = -A_1\omega^2 \cos(\omega t - \alpha)$$
$$\ddot{y}_2 = -A_2\omega^2 \cos(\omega t - \alpha)$$

Put it in the system of differential equations:

$$A_1 - \alpha_{11}^* \cdot mA_1\omega^2 - \alpha_{12}^* \cdot mA_2\omega^2 = 0$$
$$A_2 - \alpha_{21}^* \cdot mA_1\omega^2 - \alpha_{22}^* \cdot mA_2\omega^2 = 0$$

$$A_1(1 - \alpha_{11}^* \cdot m\omega^2) + A_2 \cdot (-\alpha_{12}^* \cdot m\omega^2) = 0$$
$$A_1 \cdot (-\alpha_{21}^* \cdot m\omega^2) + A_2 \cdot (1 - \alpha_{22}^* \cdot m\omega^2) = 0$$

$$\Delta(\omega^2) = \begin{vmatrix} 1 - \alpha_{11}^* \cdot m\omega^2 & -\alpha_{12}^* \cdot m\omega^2 \\ -\alpha_{21}^* \cdot m\omega^2 & 1 - \alpha_{22}^* \cdot m\omega^2 \end{vmatrix} = 0$$

$$(1 - \alpha_{11}^* \cdot m\omega^2) \cdot (1 - \alpha_{22}^* \cdot m\omega^2) - \alpha_{12}^{*\,2} \cdot m^2\omega^4 = 0$$

$$(\alpha_{11}^* \cdot \alpha_{22}^* \cdot m^2 - \alpha_{12}^{*\,2} m^2) \cdot \omega^4 - (\alpha_{22}^* m + \alpha_{11}^* \cdot m)\omega^2 + 1 = 0$$

Now, it is necessary to estimate the influence coefficients $\alpha_{11}^*;\alpha_{22}^*$ and $\alpha_{12}^*$.

$\alpha_{11}^*$ is the bilge in point ① when the unit force acts upon the beam in point ① (Figure 8.18.) It is needed to separate the system in point *C* and we will have two beams in Figure 8.19. The force $F_c$ is in point *C*. The bilge in point *C* of beams *AB* and *CD* is equal:

$$f_C^{AB} = f_C^{CD}$$
$$f_C^{AB} = \alpha_{C1}^{AB} \cdot 1 + \alpha_{CC}^{AB} \cdot (-F_C)$$
$$f_C^{CD} = \alpha_{CC}^{CD} \cdot F_C$$
$$\alpha_{C1}^{AB} \cdot 1 + \alpha_{CC}^{AB} \cdot (-F_C) = \alpha_{CC}^{CD} \cdot F_C$$

$$\alpha_{C1}^{AB} = \alpha_{CC}^{CD} \cdot F_C + \alpha_{CC}^{AB} \cdot F_C$$

$$F_C = \frac{\alpha_{C1}^{AB}}{\alpha_{CC}^{CD} + \alpha_{CC}^{AB}}$$

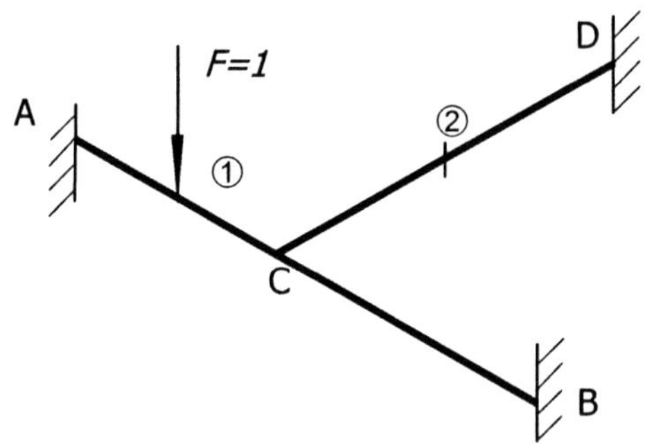


*Figure 8.18.*

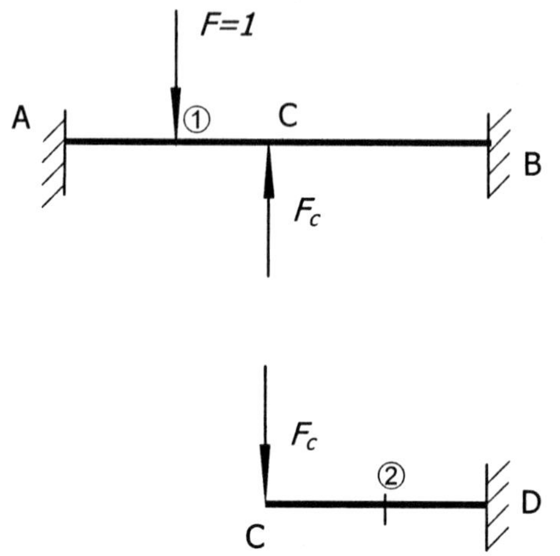


*Figure 8.19.*

The bilge of beam AB in point ① (Figure 8.18.):

$$\alpha_{11}^{*} = \alpha_{11}^{AB} \cdot 1 + \alpha_{1C}^{AB} \cdot (-F_c) = \alpha_{11}^{AB} - \alpha_{1C}^{AB} \frac{\alpha_{C1}^{AB}}{\alpha_{CC}^{CD} + \alpha_{CC}^{AB}} =$$

$$= \alpha_{11}^{*} = \alpha_{11}^{AB} - \frac{\left(\alpha_{C1}^{AB}\right)^2}{\alpha_{CC}^{CD} + \alpha_{CC}^{AB}}$$

So, it is needed to calculate the influence coefficients $\alpha_{11}^{AB}$, $\alpha_{CC}^{AB}$, $\alpha_{1C}^{AB}$, $\alpha_{CC}^{CD}$, and put it in the last expression. The coefficient $\alpha_{11}^{AB}$ is calculated using Figure 8.20. and Table 5. of appendix (3rd row):

$$\alpha_{11}^{AB}=\frac{(4l)^3}{3EI}\cdot\left(\frac{l}{4l}\right)^3\cdot\left(\frac{3l}{4l}\right)^3$$

$$\alpha_{11}^{AB}=\frac{9}{64}\frac{l^3}{EI}$$

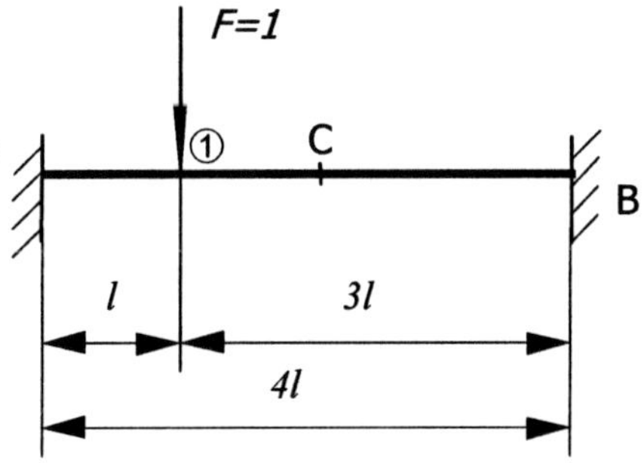


*Figure 8.20.*

For $\alpha_{CC}^{AB}$ we can use Table 5. - 3rd row and Figure 8.21:

$$\alpha_{CC}^{AB}=\frac{(4l)^3}{3EI}\cdot\left(\frac{2l}{4l}\right)^3\cdot\left(\frac{2l}{4l}\right)^3$$

$$\alpha_{CC}^{AB}=\frac{1}{3}\frac{l^3}{EI} \quad \ldots\ldots\ldots\ldots\ldots\ldots\ldots\ldots (8.2.5.)$$

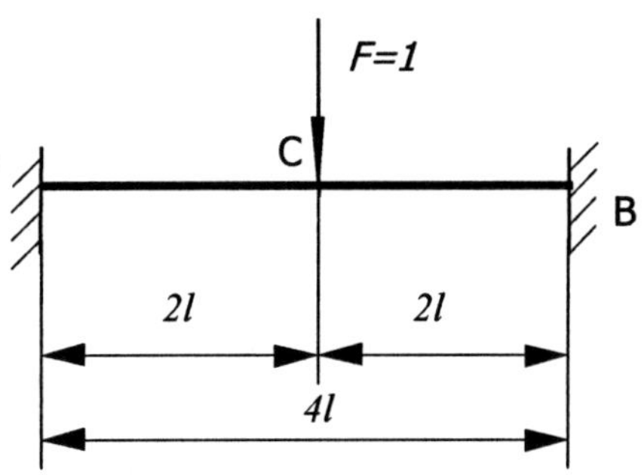


*Figure 8.21.*

Table 5. - 1st row and Figure 8.22. is used for $\alpha_{1C}^{AB}$:

$$\alpha_{1C}^{AB} = \frac{(4l)^3}{6EI} \cdot \left(\frac{2l}{4l}\right)^2 \cdot \left(\frac{l}{4l}\right)^2 \left[3\frac{2l}{4l} - \left(\frac{2l}{4l} + 3 \cdot \frac{2l}{4l}\right)\frac{l}{4l}\right]$$

$$\alpha_{1C}^{AB} = \frac{1}{6}\frac{l^3}{EI} \qquad \text{(8.2.6.)}$$

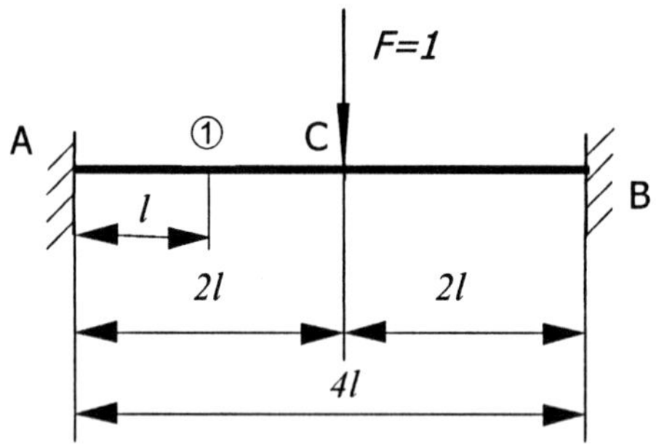


*Figure 8.22.*

$\alpha_{CC}^{CD}$ is estimated using Table 2. - 3rd row and Figure 8.23:

$$\alpha_{CC}^{CD} = \frac{(2l)^3}{3EI}$$

$$\alpha_{CC}^{CD} = \frac{8}{3}\frac{l^3}{EI} \qquad \text{(8.2.7.)}$$

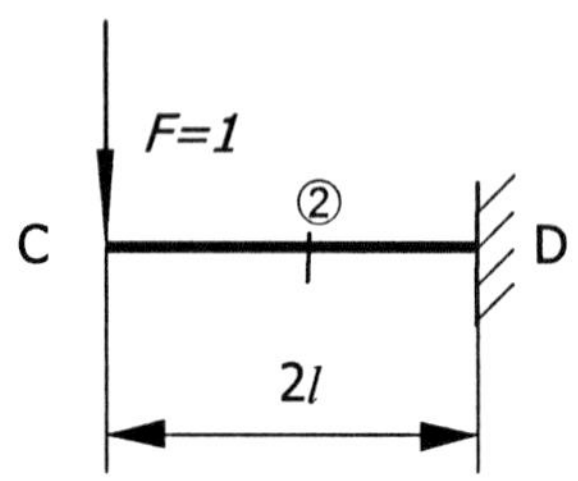


*Figure 8.23.*

The coefficient $\alpha_{11}^*$:

$$\alpha_{11}^* = \alpha_{11}^{AB} - \frac{\left(\alpha_{C1}^{AB}\right)^2}{\alpha_{CC}^{CD} + \alpha_{CC}^{AB}} = \frac{9}{64}\frac{l^3}{EI} - \frac{\left(\frac{1}{6}\frac{l^3}{EI}\right)^2}{\frac{1}{3}\frac{l^3}{EI} + \frac{8}{3}\frac{l^3}{EI}} = 0{,}0851\frac{l^3}{EI}$$

The procedure for giving $\alpha_{22}^{*}$ is the same. The bilge in point *C* of the beams *AB* and *CD* is equal:

$f_C^{AB} = f_C^{CD}$

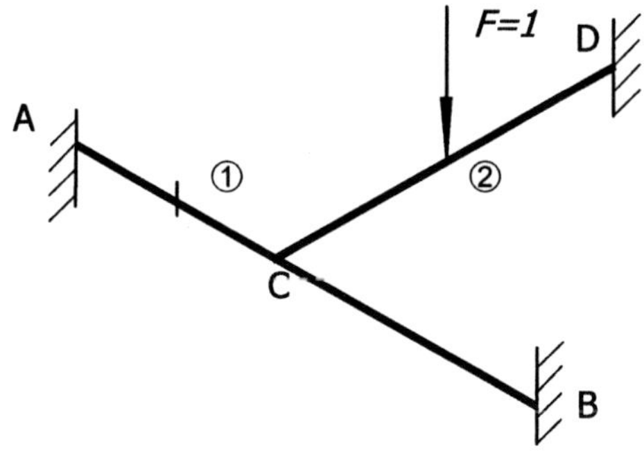


*Figure 8.24.*

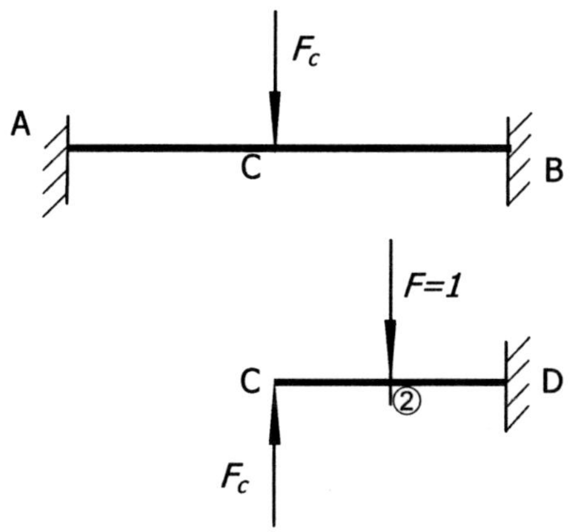


*Figure 8.25.*

$$f_C^{AB} = \alpha_{CC}^{AB} \cdot F_C$$
$$f_C^{CD} = \alpha_{CC}^{CD} \cdot (-F_C) + \alpha_{C2}^{CD} \cdot 1$$
$$\alpha_{CC}^{AB} \cdot F_C = \alpha_{CC}^{CD} \cdot (-F_C) + \alpha_{C2}^{CD} \cdot 1$$
$$F_C \cdot (\alpha_{CC}^{AB} + \alpha_{CC}^{CD}) = \alpha_{C2}^{CD}$$
$$F_C = \frac{\alpha_{C2}^{CD}}{\alpha_{CC}^{AB} + \alpha_{CC}^{CD}}$$

The bilge of beam *CD* in point ② (Figure 8.25.):

$$\alpha_{22}^{*} = \alpha_{22}^{CD} \cdot 1 + \alpha_{2C}^{CD} \cdot (-F_c) = \alpha_{22}^{CD} - \frac{\left(\alpha_{2C}^{CD}\right)^2}{\alpha_{CC}^{AB} + \alpha_{CC}^{CD}}$$

It is necessary to calculate: $\alpha_{22}^{CD}$, $\alpha_{2C}^{CD}$, $\alpha_{CC}^{AB}$ and $\alpha_{CC}^{CD}$.

$\alpha_{22}^{CD} = \frac{l^3}{3EI} = \frac{1}{3} \cdot \frac{l^3}{EI}$ - table 2. - 3rd row and Figure 8.25.

$\alpha_{2C}^{CD} = \frac{(2l)^3}{6EI} \cdot \left[ 3 \frac{2l}{2l} \left( \frac{l}{2l} \right)^2 - \left( \frac{l}{2l} \right)^3 \right]$ - table 2. - 1st row and Figure 8.26.

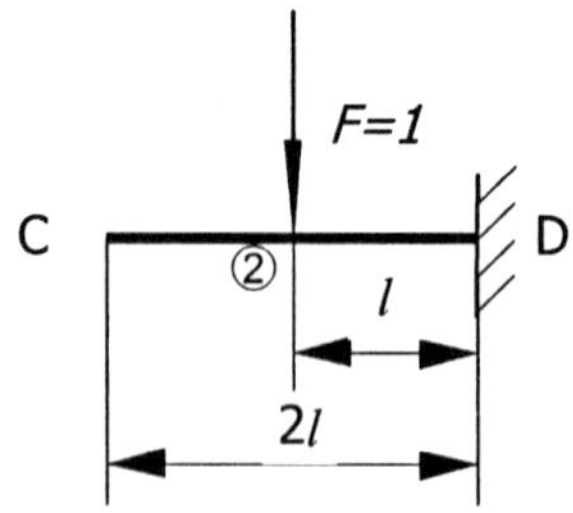


*Figure 8.26.*

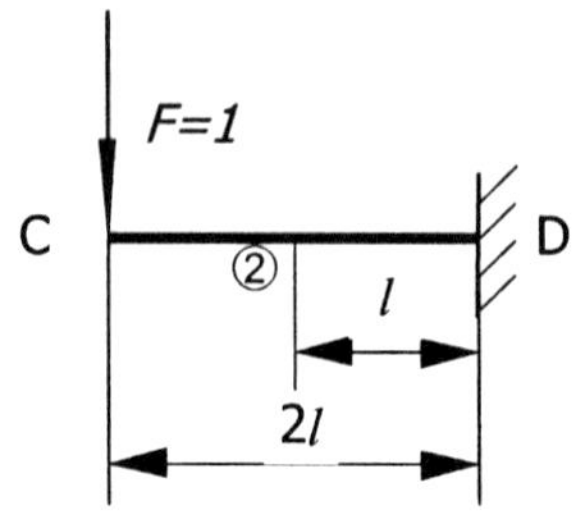


*Figure 8.27.*

$$\alpha_{2C}^{CD} = \frac{(2l)^3}{6EI} \cdot \left( 3 \frac{1}{4} - \frac{1}{8} \right) = \frac{5}{6} \frac{l^3}{EI} \qquad (8.2.8.)$$

$\alpha_{CC}^{AB}$ and $\alpha_{CC}^{CD}$ are calculated (see (8.2.5.) and (8.2.7.)).

$$\alpha_{22}^{*} = \alpha_{22}^{CD} - \frac{\left(\alpha_{2C}^{CD}\right)^2}{\alpha_{CC}^{AB} + \alpha_{CC}^{CD}} = 0{,}33 \frac{l^3}{EI} - \frac{\left( \frac{5}{6} \cdot \frac{l^3}{EI} \right)}{\frac{1}{3} \frac{l^3}{EI} + \frac{8}{3} \frac{l^3}{EI}} = 0{,}0556 \frac{l^3}{EI}$$

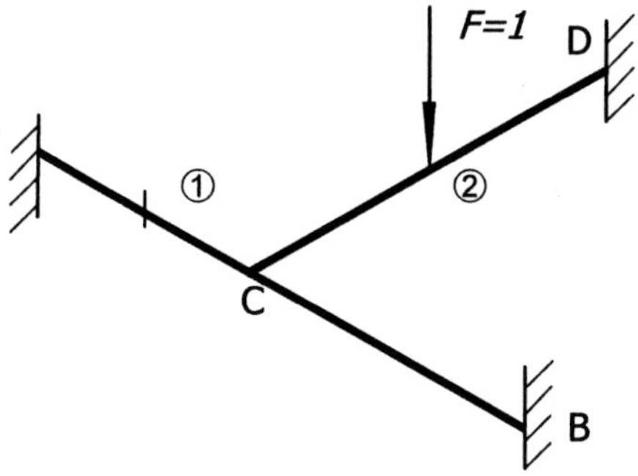


*Figure 8.28.*

Also, procedure for calculating $\alpha_{12}^*$ is the same. The bilges in point C of beams AB and CD:

$$f_C^{AB} = f_C^{CD}$$

$$f_C^{AB} = \alpha_{CC}^{AB} \cdot F_C$$

$$f_C^{CD} = \alpha_{CC}^{CD} \cdot (-F_C) + \alpha_{C2}^{CD} \cdot 1$$

$$\alpha_{CC}^{AB} \cdot F_C = \alpha_{CC}^{CD} \cdot (-F_C) + \alpha_{C2}^{CD} \cdot 1$$

$$F_C \cdot (\alpha_{CC}^{AB} + \alpha_{CC}^{CD}) = \alpha_{C2}^{CD}$$

$$F_C = \frac{\alpha_{C2}^{CD}}{\alpha_{CC}^{AB} + \alpha_{CC}^{CD}}$$

$$\alpha_{12}^* = \alpha_{1C}^{AB} \cdot F_c = \frac{\alpha_{1C}^{AB} \cdot \alpha_{2C}^{CD}}{\alpha_{CC}^{AB} + \alpha_{CC}^{CD}}$$

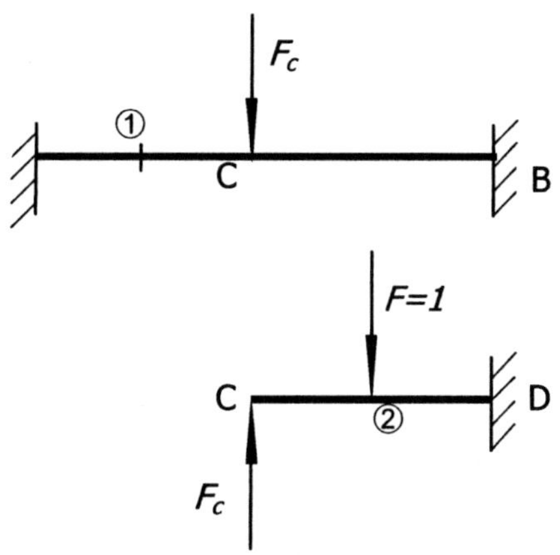


*Figure 8.29.*

The coefficients $\alpha_{1C}^{AB};\alpha_{C2}^{CD};\alpha_{CC}^{AB}$ and $\alpha_{CC}^{CD}$ are calculated (see (8.2.6.), (8.2.8.), (8.2.5.) and (8.2.7.)).

$$\alpha_{12}^{*}=\frac{\frac{1}{6}\cdot\frac{l^3}{EI}\cdot\frac{5}{6}\cdot\frac{l^3}{EI}}{\frac{1}{3}\cdot\frac{l^3}{EI}+\frac{8}{3}\cdot\frac{l^3}{EI}}=0{,}0463\cdot\frac{l^3}{EI}$$

8.5. Estimate the frequency of vibration's system in Figure 8.30. It is known: $mg$=100[N]; $c$=96[N/cm]; $l$=200[cm]; cross section of the beam is circular with diameter $d$=20[mm]; and the material is steel ($E$=2·10$^{11}$[N/m$^2$]).

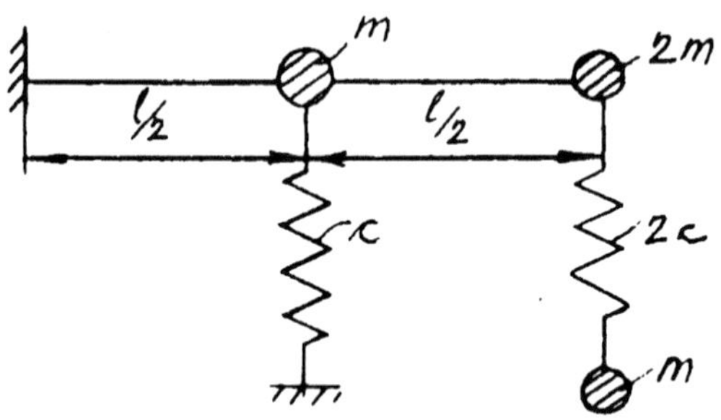


*Figure 8.30.*

Hint: Inertia's moment of the beam: $I=\dfrac{d^4\pi}{64}$

Final solution:

$\omega_1=3{,}36\left[s^{-1}\right]$;

$\omega_2=33{,}99\left[s^{-1}\right]$;

$\omega_3=43{,}68\left[s^{-1}\right]$

8.6. Estimate the frequent equation of transverse vibrations and the influence coefficients for beam in Figure 8.31.

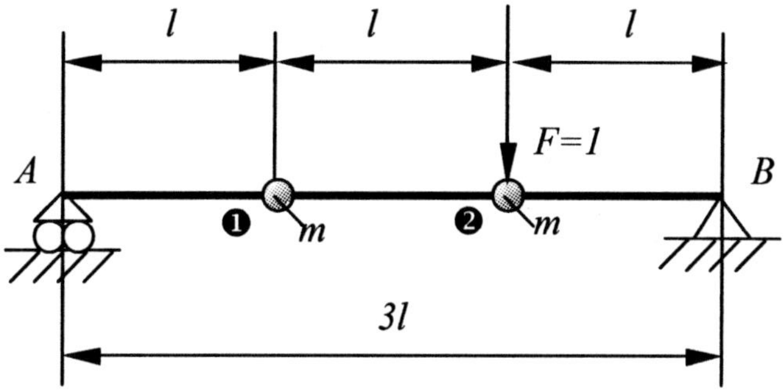


Figure 8.31.

Final solution:

$$\left(\omega^2 m\alpha_{11}-1\right)\cdot\left(\omega^2 m\alpha_{22}-1\right)-\omega^4 m^2\alpha_{12}^2=0$$

$$\alpha_{11}=\frac{(3l)^3}{3EI}\cdot\left(\frac{l}{3l}\right)^2\cdot\left(\frac{2l}{3l}\right)^2=\frac{4l^3}{9EI}$$

$$\alpha_{12}=\alpha_{21}=\frac{(3l)^3}{6EI}\cdot\left\{\frac{l}{3l}\cdot\frac{l}{3l}\left[1-\left(\frac{l}{3l}\right)^2-\left(\frac{l}{3l}\right)^2\right]\right\}=\frac{l^3}{2EI}$$

$$\alpha_{22}=\frac{(3l)^3}{3EI}\cdot\left(\frac{2l}{3l}\right)^2\cdot\left(\frac{l}{3l}\right)^2=\frac{4l^3}{9EI}$$

# CHAPTER NINE

## FORCED UNDAMPED VIBRATIONS WITH TWO DEGREES OF VIBRATION'S FREEDOM

### 9.1. Short Theory Review

If the excitative force $Q_{1\Omega}$ and $Q_{2\Omega}$ act upon vibration's system then there are forced vibrations. It is possible that $Q_{1\Omega}=0$ and $Q_{2\Omega}\neq0$ or $Q_{1\Omega}\neq0$ and $Q_{2\Omega}=0$. The excitative forces can have the form of a single harmonic function of time:

$$Q_{1\Omega} = H_1 \sin(\Omega t + \beta)$$
$$Q_{2\Omega} = H_2 \sin(\Omega t + \beta) \qquad \text{(9.1.)}$$

LaGrange's equations have the next form for forced undamped vibrations with two degrees of vibration's freedom:

$$\frac{d}{dt}\left(\frac{\partial E_k}{\partial \dot{q}_1}\right) - \frac{\partial E_k}{\partial q_1} + \frac{\partial E_p}{\partial q_1} = Q_{1\Omega}$$
$$\frac{d}{dt}\left(\frac{\partial E_k}{\partial \dot{q}_2}\right) - \frac{\partial E_k}{\partial q_2} + \frac{\partial E_p}{\partial q_2} = Q_{2\Omega} \qquad \text{(9.2.)}$$

where $E_k$ is kinetic energy, $E_p$ is potential energy, $q_1$ and $q_2$ are generalised coordinate, $Q_{1\Omega}$ and $Q_{2\Omega}$ are excitative forces. We will get a system of differential equations:

$$a_{11}\ddot{q}_1 + a_{12}\ddot{q}_2 + c_{11}q_1 + c_{12}q_2 = Q_{1\Omega}$$
$$a_{21}\ddot{q}_1 + a_{22}\ddot{q}_2 + c_{21}q_1 + c_{22}q_2 = Q_{2\Omega} \qquad \text{(9.3.)}$$

where:

$a_{11};a_{12};a_{21}$ and $a_{22}$ are the system's coefficients of inertia

$c_{11};c_{12};c_{21}$ and $c_{22}$ are the system's coefficients of rigidity.

The expressions (9.3.) are the system of inhomogeneous differential equations of the second order with constant coefficients. General solutions are:

$$q_1 = (q_1)_H + (q_1)_P; \quad q_2 = (q_2)_H + (q_2)_P \qquad \text{(9.4.)}$$

where $(q_1)_H$ and $(q_2)_H$ are homogeneous, $(q_1)_P$ and $(q_2)_P$ are the particular parts of solutions.

The homogeneous parts $(q_i)_H$ and $(q_2)_H$ are the solutions of the system (7.2.) in chapter seven, and it is possible to find $(q_i)_H$ and $(q_2)_H$ using equations (7.12.). the solutions $(q_i)_H$ and $(q_2)_H$ characterise the free vibrations of the system. The particular parts $(q_1)_P$ and $(q_2)_P$ characterize forced vibrations of the system, and they have the same form as excitative force(s) (9.1.).

$$(q_1)_P = A_1 \sin(\Omega t + \beta); \quad (q_2)_P = A_2 \sin(\Omega t + \beta) \quad \text{. . (9.5.)}$$

It is needed to find the second derivative $(\ddot{q}_1)_P$ and $(\ddot{q}_2)_P$. Put it and (9.5.) in the system (9.3.) and abbreviate the equations with sin($\Omega t+\beta$):

$$\begin{aligned} A_1 \cdot \left(-a_{11}\omega^2 + c_{11}\right) + A_2 \cdot \left(-a_{12}\omega^2 + c_{12}\right) &= H_1 \\ A_1 \cdot \left(c_{12} - a_{12} \cdot \omega^2\right) + A_2 \cdot \left(c_{22} - a_{22}\omega^2\right) &= H_2 \end{aligned} \quad \text{. . . . (9.6.)}$$

The amplitudes of excitative vibrations $A_1$ and $A_2$ are:

$$A_1 = \frac{\Delta_1}{\Delta} = \frac{\begin{vmatrix} H_1 & c_{12} - a_{12}\Omega^2 \\ H_2 & c_{22} - a_{22}\Omega^2 \end{vmatrix}}{\begin{vmatrix} c_{11} - a_{11}\Omega^2 & c_{12} - a_{12}\Omega^2 \\ c_{21} - a_{21}\Omega^2 & c_{22} - a_{22}\Omega^2 \end{vmatrix}} \quad \text{. . . . . . . .(9.7.)}$$

$$A_2 = \frac{\Delta_2}{\Delta} = \frac{\begin{vmatrix} c_{11} - a_{11}\Omega^2 & H_1 \\ c_{21} - a_{21}\Omega^2 & H_2 \end{vmatrix}}{\begin{vmatrix} c_{11} - a_{11}\Omega^2 & c_{12} - a_{12}\Omega^2 \\ c_{21} - a_{21}\Omega^2 & c_{22} - a_{22}\Omega^2 \end{vmatrix}} \quad \text{. . . . . .(9.8.)}$$

Put the amplitude $A_1$ i $A_2$ in (9.5.) and you will have the particular parts of solution. The general solutions are (9.4.).

## 9.2. Problems

9.1. The homogeneous disc has mass $M$=3m, radius $R$ and it rotates around the axis in point $O_1$. It is linked to the lever with a spring. The spring has rigidity $c$=1000[N/m]. The lever rotates around axis $O_2$ and it has mass $m$ and $2m$ on the ends (Figure 9.1.). Excitative moment m=$m_0\cos(\Omega t+\beta)$ acts upon the disc. Estimate the excitative motion of the vibration's system (particular parts of solutions), if it does small vibrations. All resistance to motion is neglected. It is given: $l = 0{,}5[m]$; $\beta = \frac{\pi}{12}[rad]$ ; $\Omega = 1\left[{rad}/{s}\right]$ ; $m = 0{,}1[kg]$ ; $m_0$=1 $[Nm]$; $R = l = 0{,}5[m]$.

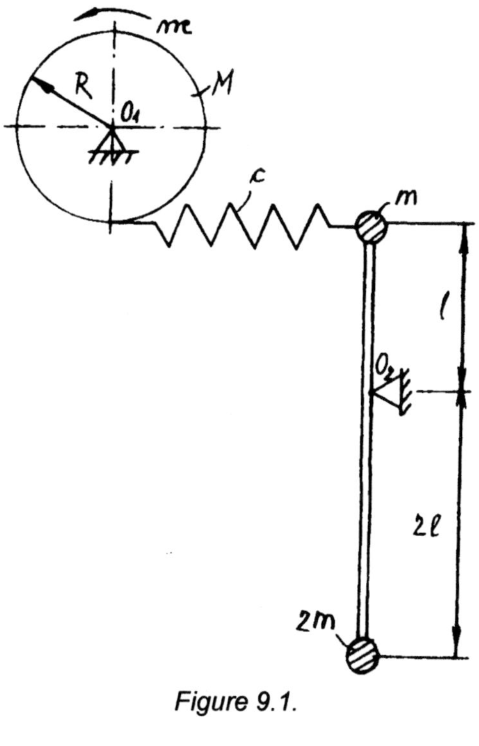


*Figure 9.1.*

Solution:

The system is in the position of equilibrium in Figure 9.1. Figure 9.2. shows the system out the position of equilibrium. The disc is displaced for angle $\varphi_1$ and the lever for $\varphi_2$. Total kinetic energy is the sum of kinetic energy of masses $m$ ($E_{km}$), $2m$ ($E_{k2m}$) and the disc ($E_{km}$).

$$E_k = E_{km} + E_{k2m} + E_{kM} = \frac{1}{2}ml^2\dot{\varphi}_2^2 + \frac{1}{2}2m4l^2\dot{\varphi}_2^2 + \frac{1}{2}J_{O1}\dot{\varphi}_1^2$$

The moment of inertia for the disc's axis in point $O_1$:

$$J_{O1} = \frac{1}{2} MR^2$$

Put it in the expression for $E_k$:

$$E_k = \frac{9}{2} ml^2 \dot{\varphi}_2^2 + \frac{3}{4} ml^2 \dot{\varphi}_1^2$$

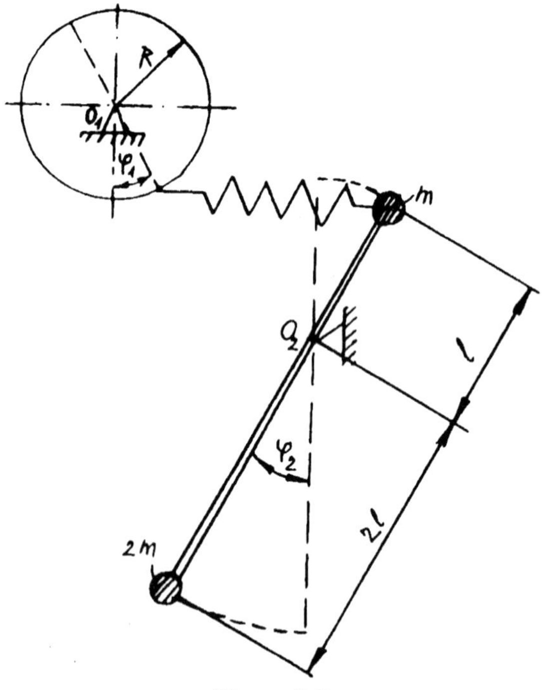


*Figure 9.2.*

Total potential energy is the sum of potential energies of masses $m$ ($E_{pm}$), $2m$ ($E_{p2m}$) and the spring ($E_{pc}$).

$$E_P = E_{pm} + E_{p2m} + E_{pc}$$

$$E_P = -mgl\frac{\varphi_2^2}{2} + \frac{4}{2} mgl\varphi_2^2 + \frac{1}{2} c(l\varphi_2 - R\varphi_1)^2$$

Set the expression in order:

$$E_p = \frac{3}{2} mgl\varphi_2^2 + \frac{1}{2} c(l\varphi_2 - l\varphi_1)^2$$

Now, put $E_k$ and $E_p$ in LaGrange's equations (9.2.):

$$2\frac{3}{4}ml^2\ddot{\varphi}_1 + 2\frac{1}{2}c(l\varphi_2 - l\varphi_1)\cdot(-l) = m_0\cos(\Omega t+\beta)$$

$$2\frac{9}{2}ml^2\ddot{\varphi}_2 + 2\frac{3}{2}mgl\varphi_2 + 2\frac{1}{2}c(l\varphi_2 - l\varphi_1)l = 0$$

When we set the system of differential equations in order, we will get:

$$\frac{3}{2}ml^2\ddot{\varphi}_1 - cl^2(\varphi_2 - \varphi_1) = m_0\cos(\Omega t+\beta)$$

. . . . . . . (9.2.1.)

$$9ml^2\ddot{\varphi}_2 + 3mgl\varphi_2 + cl^2\varphi_2 - cl^2\varphi_1 = 0$$

The particular parts of the solutions $(\varphi_1)_p$ and $(\varphi_2)_p$ have the form of the excitative moment $m=m_0\cos(\Omega t+\beta)$. So, the particular parts must have amplitude (for example $A_1$ and $A_2$) and $\cos(\Omega t+\beta)$:

$$(\varphi_1)_P = A_1\cos(\Omega t+\beta); \quad (\varphi_2)_P = A_2\cos(\Omega t+\beta)$$

$$(\ddot{\varphi}_1)_P = -A_1\Omega^2\cos(\Omega t+\beta); \quad (\ddot{\varphi}_2)_P = -A_2\Omega^2\cos(\Omega t+\beta)$$

Put it in the system (9.2.1.) and abbreviate both equations with $\cos(\Omega t+\beta)$:

$$-\frac{3}{2}ml^2A_1\Omega^2 - cl^2A_2 + cl^2A_1 = m_0$$

$$-9ml^2A_2\Omega^2 + (3mgl + cl^2)A_2 - cl^2A_1 = 0$$

We can calculate the constant (amplitude of the particular parts) $A_1$ and $A_2$ from (9.7.) and (9.8.). So, we need the determinants $\Delta$, $\Delta_1$ and $\Delta_2$:

$$\Delta = \begin{vmatrix} cl^2 - \frac{3}{2}ml^2\Omega^2 & -cl^2 \\ -cl^2 & 3mgl + cl^2 - 9ml^2\Omega^2 \end{vmatrix};$$

$$\Delta_1 = \begin{vmatrix} m_0 & -cl^2 \\ 0 & 3mgl + cl^2 - 9ml^2\Omega^2 \end{vmatrix};$$

$$\Delta_2 = \begin{vmatrix} cl^2 - \frac{3}{2}ml^2\Omega^2 & \mathrm{m}_0 \\ -cl^2 & 0 \end{vmatrix}.$$

It is necessary to develop the determinants Δ, $\Delta_1$ and $\Delta_2$, put the values of *c, l, m, Ω,* g=9,81[m/s$^2$], $m_0$ and we will get:

$$\Delta = 139{,}116\left[N^2m^2\right]; \quad \Delta_1 = 250{,}5715\left[N^2m^2\right];$$

$$\Delta_2 = 250\left[N^2m^2\right]$$

The amplitudes:

$$A_1 = \frac{\Delta_1}{\Delta} = 1{,}801[rad]; \quad A_2 = \frac{\Delta_2}{\Delta} = 1{,}797[rad]$$

The excitative motion of vibration's system:

$$\varphi_{1p} = 1{,}801\cos\left(1\cdot t + \frac{\pi}{12}\right)[rad, s];$$

$$\varphi_{2p} = 1{,}797\cos\left(1\cdot t + \frac{\pi}{12}\right)[rad, s]$$

9.2. An engine has weight $G_1$=100$[kN]$ and it is situated on elastic ground. The rigidity of the ground is $c_1$. The excitative force $F_\Omega = F_0 \sin\Omega t[N,s]$ acts upon the engine. The system does the vertical forced vibrations. Amplitude of the excitative force is $F_0 = 10[kN]$. The resonance appears when the angle speed of the engine's shaft is $\Omega = 100\left[s^{-1}\right]$. The vibration's absorber is built on the engine in the form of weight $G_2$ with the springs (rigidity $c_2$) for reducing the vibrations (Figure 9.3.). How much is the weight $G_2$ and equivalent rigidity of the springs $c_2$, if the amplitude of the forced vibrations is $A_1$=0[mm] and amplitude of the weight's vibrations is $A_2$=2[mm].

Solution:

**This is the general example how can we reduce the vibrations.** It is necessary to add the vibration's absorber like in Figure 9.3.

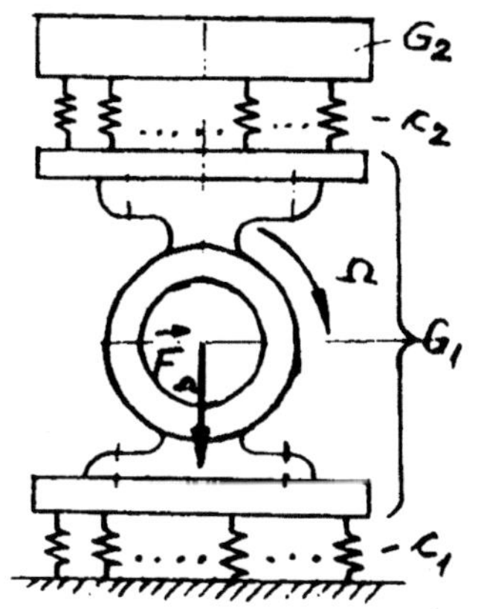


*Figure 9.3.*

Figure 9.4. shows the mechanical model of the system. The forces in the springs are: $F_{c1}=c_1y_1$ ; $F_{c2}=c_2(y_2-y_1)$. Newton's second law for the engine:

$$\frac{G_1}{g}\ddot{y}_1 = F_{\Omega} - c_1 y_1 + c_2(y_2 - y_1)$$

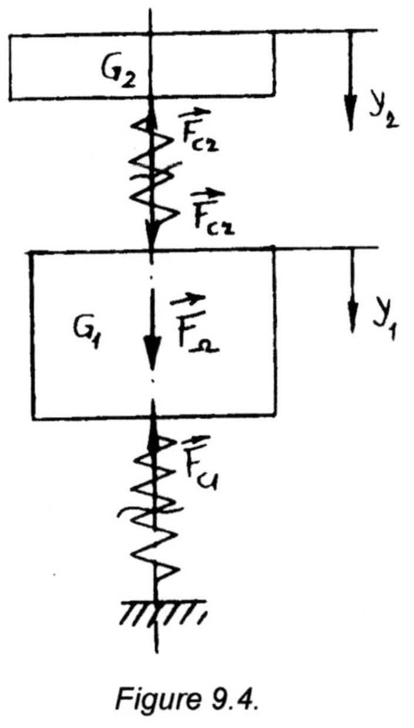


*Figure 9.4.*

The Newton's second law for the weight $G_2$:

$$\frac{G_2}{g}\ddot{y}_2 = -c_2(y_2 - y_1)$$

The last two equations are the system of differential equations:

$$\frac{G_1}{g}\ddot{y}_1 = F_\Omega - c_1 y_1 + c_2(y_2 - y_1)$$

$$\frac{G_2}{g}\ddot{y}_2 = -c_2(y_2 - y_1)$$

$$\frac{G_1}{g}\ddot{y}_1 + c_1 y_1 - c_2(y_2 - y_1) = F_0 \sin\Omega t$$

$$\frac{G_2}{g}\ddot{y}_2 + c_2(y_2 - y_1) = 0$$

$$A_1\left(c_1 + c_2 - \frac{G_1}{g}\Omega^2\right) + A_2(-c_2) = F_0$$

$$A_1(-c_2) + A_2\left(c_2 - \frac{G_2}{g}\Omega^2\right) = 0$$

$$A_2 = \frac{\Delta_2}{\Delta} = \frac{\begin{vmatrix} c_1 + c_2 - \frac{G_1}{g}\Omega^2 & F_0 \\ -c_2 & 0 \end{vmatrix}}{\Delta} = \frac{F_0 c_2}{\left(c_1 + c_2 - \frac{G_2}{g}\Omega^2\right)\left(c_2 - \frac{G_2}{g}\Omega^2\right) - c_2^2} \quad ..(9.2.2.)$$

$$A_1 = \frac{\Delta_1}{\Delta} = \frac{\begin{vmatrix} F_0 & -c_2 \\ 0 & c_2 - \frac{G_2}{g}\Omega^2 \end{vmatrix}}{\Delta} \quad \text{. . . . . . . . . . . . . . . . . . (9.2.3.)}$$

The amplitude of forced vibrations $A_1$=0 $\Rightarrow \frac{\Delta_1}{\Delta} \Rightarrow \Delta_1$=0

$$F_0 \cdot \left(c_2 - \frac{G_2}{g}\Omega^2\right) = 0$$

$F_0 \neq 0$, and $c_2 - \frac{G_2}{g}\Omega^2$ must be 0.

$$c_2 - \frac{G_2}{g}\Omega^2 = 0 \quad \ldots\ldots\ldots\ldots\ldots\ldots (9.2.4.)$$

Put it in (9.2.2.):

$$A_2 = \frac{F_0 c_2}{-c_2^2} \quad \Leftrightarrow \quad A_2 = -\frac{F_0}{c_2}$$

Amplitude of the weight's vibrations $A_2$=2[mm]. We must take the absolute value, because the amplitude can't be less than 0.

$$A_2 = \left|\frac{F_0 c_2}{-c_2^2}\right| = \left|\frac{F_0}{-c_2}\right| = \frac{F_0}{c_2} \quad \Rightarrow$$

$$\Rightarrow \quad c_2 = \frac{F_0}{A_2} = \frac{10[kN]}{2[mm]} = 5\left[\frac{kN}{mm}\right] = 50\left[\frac{kN}{cm}\right]$$

Put the value of $c_2$ in (9.2.4.):

$$G_2 = \frac{c_2 g}{\Omega^2} = \frac{50\left[\frac{kN}{cm}\right]\cdot 981\left[\frac{cm}{s^2}\right]}{100^2\left[s^{-2}\right]} = 4{,}9[kN]$$

If the weight $G_2$=4,9[kN] and rigidity $c_2$=50[kN/cm], the amplitude of the engine $A_1$=0.

9.3. A bar AB has mass 2m and it moves without sliding on two discs (Figure 9.5.). The discs have mass *m*, radius *r* and they rotate around the individual axes. Unextensible rope with concentrated mass *M* is hooked up in point C of the bar. The excitative force $F_\Omega = H\cos\Omega t$ acts upon the bar.

Estimate:

a) the amplitudes of forced vibrations,

b) the law of forced disc's vibrations.

Solution:

Kinetic energy (Figure 9.6.): $E_k = 2E_{km} + E_{kM} + E_k^{AB}$

$$E_{km} = \frac{1}{2} J_m \dot{\varphi}_1^2 = \frac{1}{2}\frac{1}{2} mr^2 \dot{\varphi}_1^2 = \frac{mr^2}{4} \dot{\varphi}_1^2$$

where $J_m = \frac{1}{2} mr^2$ is inertia's moment of the disc for the individual axis.

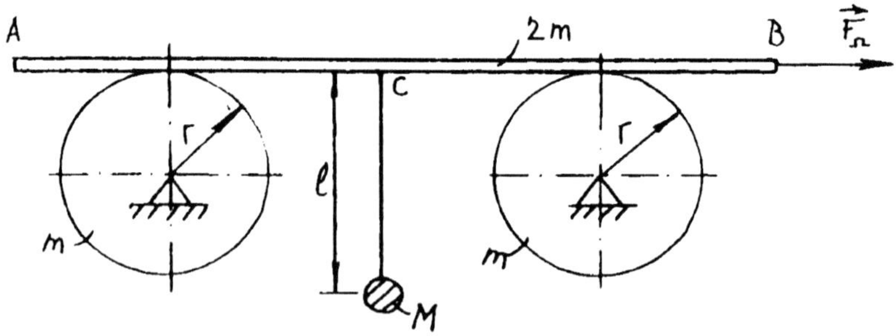


*Figure 9.5.*

$$E_{kM} = \frac{1}{2} M v_M^2$$

It is necessary to find $v_m$:

$\vec{v}_M = \vec{v}_C + \vec{v}_M^C$ - this is the vectorial equation. For small vibrations it is possible to approximate it with algebraic equation:

$v_M = v_C + v_M^C$,

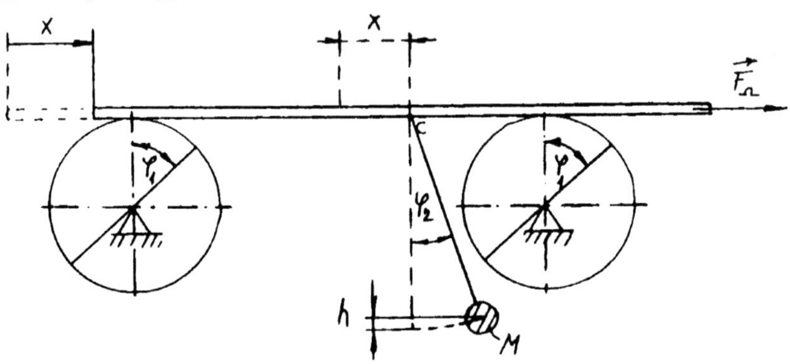


*Figure 9.6.*

The speed of point C: $v_c = \frac{dx}{dt} = \dot{x}$

The speed of point *M* round point *C*: $v_M^C = l \cdot \frac{d\varphi_2}{dt} = l \cdot \dot{\varphi}_2$.

So, the speed of point *M*:

$$v_M = \dot{x} + l\dot{\varphi}_2 \qquad \text{. . . . . . . . . . . . . . . . . . (9.2.5.)}$$

Put (9.2.5.) in the expression of $E_{kM}$:

$$E_{kM} = \frac{1}{2} M\left(\dot{x} + l\dot{\varphi}_2\right)^2$$

Relation between the coordinates *x* and $\varphi_1$: $x = r \cdot \varphi_1$

$$\frac{dx}{dt} = r \cdot \frac{d\varphi_1}{dt}$$

$$\dot{x} = r \cdot \dot{\varphi}_1$$

Now, the expression of $E_{kM}$:

$$E_{kM} = \frac{1}{2} M\left(r\dot{\varphi}_1 + l\dot{\varphi}_2\right)^2$$

The kinetic energy of the bar *AB*:

$$E_k^{AB} = \frac{1}{2} 2m\dot{x}^2 = \frac{1}{2} 2m\left(r\dot{\varphi}_1\right)^2 = mr^2\dot{\varphi}_1^2$$

$$E_k = \frac{mr^2}{2}\dot{\varphi}_1^2 + \frac{1}{2} M\left(r\dot{\varphi}_1 + l\dot{\varphi}_2\right)^2 + mr^2\dot{\varphi}_1^2$$

The potential energy:

$$E_p = E_{pM} = Mgh = Mg\left(l - l\cos\varphi_2\right)$$

For small vibrations $\cos\varphi \approx 1 - \frac{\varphi^2}{2}$

$$E_p = Mgl\left(1 - 1 + \frac{\varphi_2^2}{2}\right) = Mgl\frac{\varphi_2^2}{2}$$

Put $E_k$ and $E_p$ in the LaGrange's equations (9.2.):

$$\frac{d}{dt}\left(\frac{\partial E_k}{\partial \dot{q}_1}\right) - \frac{\partial E_k}{\partial q_1} + \frac{\partial E_p}{\partial q_1} = Q_{1\Omega}$$

$$\frac{d}{dt}\left(\frac{\partial E_k}{\partial \dot{q}_2}\right) - \frac{\partial E_k}{\partial q_2} + \frac{\partial E_p}{\partial q_2} = Q_{2\Omega}$$

$$mr^2\ddot{\varphi}_1 + Mr^2\ddot{\varphi}_1 + Mlr\ddot{\varphi}_2 + 2mr^2\ddot{\varphi}_1 = H\cos\Omega t$$
$$M(r\ddot{\varphi}_1 + l\ddot{\varphi}_2)l + Mgl\varphi_2 = 0$$

$$mr^2\ddot{\varphi}_1 + Mr^2\ddot{\varphi}_1 + Mlr\ddot{\varphi}_2 + 2mr^2\ddot{\varphi}_1 = H\cos\Omega t$$
$$Mrl\ddot{\varphi}_1 + Ml^2\ddot{\varphi}_2 + Mgl\varphi_2 = 0$$

Particular parts of the solutions have the form of excitative force $F_\Omega$:

$$\varphi_1 = A_1\cos\Omega t; \varphi_2 = A_2\cos\Omega t;$$
$$\ddot{\varphi}_1 = -\Omega^2 A_1\cos\Omega t; \ddot{\varphi}_2 = -\Omega^2 A_2\cos\Omega t$$

Put it in the system of differential equations and abbreviate it with $\cos\Omega t$:

$$-mr^2\Omega^2 A_1 + Mr^2(-\Omega^2)A_1 + Mlr(-\Omega^2)A_2 + 2mr^2(-\Omega^2)A_1 = H$$
$$Mrl(-\Omega^2)A_1 + Ml^2(-\Omega^2)A_2 + MglA_2 = 0$$

$$(-mr^2\Omega^2 - Mr^2\Omega^2 - 2mr^2\Omega^2)A_1 - Mlr\Omega^2 A_2 = H$$
$$-Mlr\Omega^2 A_1 + (Mgl - Ml^2\Omega^2)A_2 = 0$$

$$(-3mr^2 - Mr^2)\Omega^2 A_1 - Mlr\Omega^2 A_2 = H$$
$$-Mlr\Omega^2 A_1 + (g - l\Omega^2)MlA_2 = 0$$

The determinants are:

$$\Delta = \begin{vmatrix} -3m\ r^2\Omega^2 - Mr^2\Omega^2 & -lr\Omega^2 M \\ -Mlr\Omega^2 & gMl - l^2 M\Omega^2 \end{vmatrix} =$$
$$= 3mMr^2l^2\Omega^4 - (3mMr^2l\cdot g + r^2gM^2l)\Omega^2$$

$$\Delta_1 = \begin{vmatrix} H & -lr\Omega^2 M \\ 0 & gMl - l^2 M\Omega^2 \end{vmatrix} = H(gMl - l^2 M\Omega^2)$$

$$\Delta_2 = \begin{vmatrix} -3mr^2\Omega^2 - Mr^2\Omega^2 & H \\ -Mlr\Omega^2 & 0 \end{vmatrix} = HMlr\Omega^2$$

a) The amplitudes of forced vibrations are:

$$A_1 = \frac{\Delta_1}{\Delta} = \frac{H(Mgl - Ml^2\Omega^2)}{(-Mr^2\Omega^2 - 3mr^2\Omega^2)(Mgl - Ml^2\Omega^2) - (-Mlr\Omega^2)^2}$$

$$A_2 = \frac{\Delta_2}{\Delta} = \frac{HMrl\Omega}{\left(-Mr^2\Omega^2 - 3mr^2\Omega^2\right)\left(Mgl - Ml^2\Omega^2\right) - \left(-Mlr\Omega^2\right)^2}$$

b) The law of forced disc's vibrations:

$$\varphi_1 = \frac{H\left(Mgl - Ml^2\Omega^2\right)}{\left(-Mr^2\Omega^2 - 3mr^2\Omega^2\right)\left(Mgl - Ml^2\Omega^2\right) - \left(-Mlr\Omega^2\right)^2}\cos\Omega t$$

9.4. The vibration system is composed of beam *AB*, the spring with rigidity $c_1$, the mass $2m$, the spring of rigidity $c_2$ and mass $m$ (Figure 8.7.). The excitative force $F_\Omega = F_0 \cdot \sin\Omega t$ acts upon the mass $m$. Estimate the amplitudes of forced vibrations. It is known: $m$=50[kg]; $c_1$=5000[N/m]; $l$=2[m]; $F_0$=500[N]; $\Omega$=4[s$^{-1}$]; $E=2\cdot10^{11}$[N/m$^2$]; $I=8\cdot10^{-6}$[m$^4$].

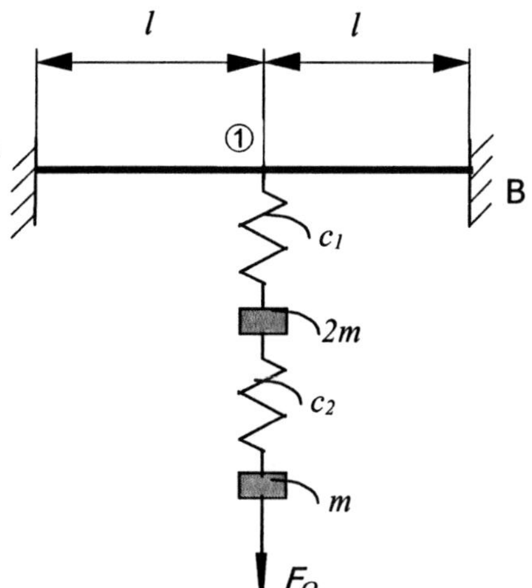

*Figure 9.7.*

Solution:

It is possible to take the beam like a flexible spring. The rigidity of the flexible spring is $c_f = \frac{1}{\alpha_{ii}}$, where $\alpha_{ii}$ is the influence coefficient (see chapter 8). So, the rigidity of the beam AB in point ①:

$$c_b=\frac{1}{\alpha_{11}}.$$

The beam *AB* and the spring of rigidity $c_1$ are a series linked in point ①, and it is possible to replace them with one (Figure 9.8.), which has equivalent rigidity $c_{ekv}$:

$$\frac{1}{c_{ekv}}=\frac{1}{c_b}+\frac{1}{c_1}.$$

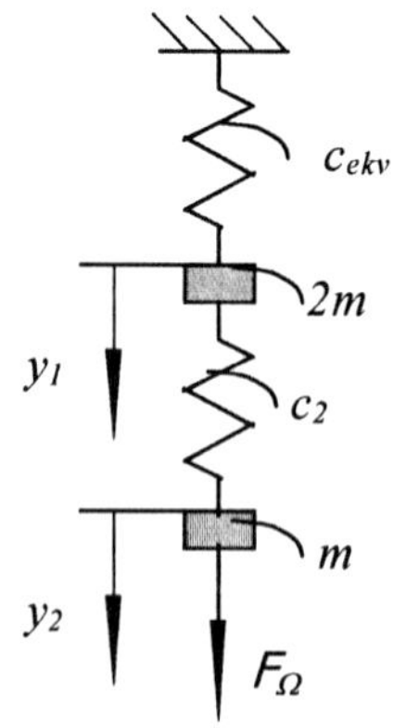


*Figure 9.8.*

It is necessary to estimate the influence coefficient $\alpha_{11}$ using Figure 9.9. and table 5. (3rd row) of appendix:

$$\alpha_{11}=\frac{(2l)^3}{3EI}\cdot\left(\frac{l}{2l}\right)^3\cdot\left(\frac{l}{2l}\right)^3=\frac{1}{24}\cdot\frac{l^3}{EI}=$$

$$=\frac{1}{24}\cdot\frac{2^3\left[m^3\right]}{2\cdot10^{11}\left[\frac{N}{m^2}\right]\cdot8\cdot10^{-6}\left[m^4\right]}=0{,}208\cdot10^{-6}\left[\frac{m}{N}\right]$$

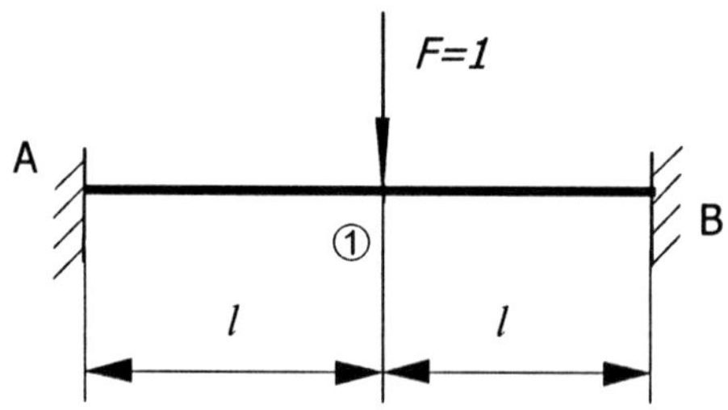


*Figure 9.9.*

The equivalent rigidity is:

$$\frac{1}{c_{ekv}} = \frac{1}{\frac{1}{\alpha_{11}}} + \frac{1}{c_1}$$

$$\frac{1}{c_{ekv}} = \alpha_{11} + \frac{1}{c_1}$$

$$\frac{1}{c_{ekv}} = \frac{\alpha_{11} c_1 + 1}{c_1}$$

$$c_{ekv} = \frac{c_1}{\alpha_{11} c_1 + 1} = \frac{25000\left[\frac{N}{m^2}\right]}{0{,}208 \cdot 10^{-6}\left[\frac{m}{N}\right] \cdot 25000\left[\frac{N}{m}\right] + 1} = 24870{,}67\left[\frac{N}{m}\right]$$

The kinetic energy of the system:

$$E_k = \frac{1}{2} \cdot 2m \cdot y_1 + \frac{1}{2} \cdot m \cdot \dot{y}_2^2$$

The potential energy:

$$E_p = \frac{1}{2} \cdot c_{ekv} \cdot y_1^2 + \frac{1}{2} \cdot c_2 \cdot (y_2 - y_1)^2$$

LaGrange's equation:

$$\frac{d}{dt}\left(\frac{\partial E_k}{\partial \dot{y}_i}\right) - \frac{\partial E_k}{\partial y_i} + \frac{\partial E_p}{\partial y_i} = Q_i; \quad i=1,2$$

Put $E_k$ and $E_p$ in the LaGrange's equations:

$$2m\ddot{y}_1 + c_{ekv} \cdot y_1 + c_2 (y_2 - y_1) \cdot (-1) = 0$$

$$m\ddot{y}_2 + c_2 (y_2 - y_1) = F_0 \cdot \sin \Omega t$$

$$y_1 = A_1 \sin \Omega t; \qquad y_2 = A_2 \sin \Omega t$$

$$\ddot{y}_1 = -\Omega^2 A_1 \sin \Omega t; \quad \ddot{y}_2 = -\Omega^2 A_2 \sin \Omega t$$

$$-2m\Omega^2 A_1 + c_{ekv} \cdot A_1 + c_2 (A_2 - A_1) \cdot (-1) = 0$$

$$-m\Omega^2 A_2 + c_2 (A_2 - A_1) = F_0$$

$$(c_{ekv} - c_2 - 2m\Omega^2) A_1 + c_2 A_2 = 0$$

$$-c_2 A_1 + (c_2 - m\Omega^2) A_2 = F_0$$

The determinants $\Delta$, $\Delta_1$ and $\Delta_2$ are:

$$\Delta=\begin{vmatrix} c_{ekv}-c_2-2m\Omega^2 & c_2 \\ -c_2 & c_2-m\Omega^2 \end{vmatrix}=$$

$$(c_{ekv}-c_2-2m\Omega^2)(c_2-m\Omega^2)+c_2^2$$

$$\Delta_1=\begin{vmatrix} 0 & c_2 \\ F_0 & c_2-m\Omega^2 \end{vmatrix}=-F_0c_2$$

$$\Delta_2=\begin{vmatrix} c_{ekv}-c_2-2m\Omega^2 & 0 \\ -c_2 & F_0 \end{vmatrix}=F_0(c_{ekv}-c_2-2m\Omega^2)$$

$\Delta = (24870{,}67[\mathrm{N/m}] - 5000[\mathrm{N/m}] - 2\cdot 50[\mathrm{kg}] \cdot 4\,[\mathrm{s}^{-2}]) \cdot$
$\cdot\,(5000[\mathrm{N/m}] - 50[\mathrm{kg}] \cdot 4\,[\mathrm{s}^{-2}]) + 5000^2[\mathrm{N}^2/\mathrm{m}^2] =$
$=1{,}089 \cdot 10^8\,[\mathrm{N}^2/\mathrm{m}^2]$

$\Delta_1 = -500[\mathrm{N}] \cdot 5000[\mathrm{N/m}] = -\,2{,}5\cdot 10^6[\mathrm{N}^2/\mathrm{m}^2]$

$\Delta_2 = 500[\mathrm{N}] \cdot (24870{,}67[\mathrm{N/m}] - 5000[\mathrm{N/m}] - 2\cdot 50[\mathrm{kg}] \cdot 4\,[\mathrm{s}^{-2}])=$
$= 9{,}735\cdot 10^6[\mathrm{N}^2/\mathrm{m}^2]$

The amplitudes of forced vibrations:

$$A_1=\frac{\Delta_1}{\Delta}=-2{,}296\cdot 10^{-2}\,[\mathrm{m}]$$

$$A_2=\frac{\Delta_2}{\Delta}=8{,}939\cdot 10^{-2}\,[\mathrm{m}]$$

9.5. There is the vibration's system in Figure 9.10, which is composed of a physical and mathematical pendulum, the spring and concentrated mass *m*. The physical pendulum has mass *M* and the mathematical *m*. The excitative moment m acts upon the physical pendulum. Estimate the distance *z* (from point $O_1$ to concentrated mass *m* in Figure 9.10.), if the amplitude of mathematical pendulum forced vibrations is $A_{mp}$=0,1[m]. It is known: $l=1[m]$; *m*=100[kg]; *c*=981[N/m]; $\Omega$=7[s$^{-1}$]; *L*=$2\cdot l$=2·1=2[m]; $m_0$=5880[Nm]; m=$m_0\cos\Omega t$; *M*=3·*m*=3·100=300[kg]

Final solution:

*z*=1,04[m]

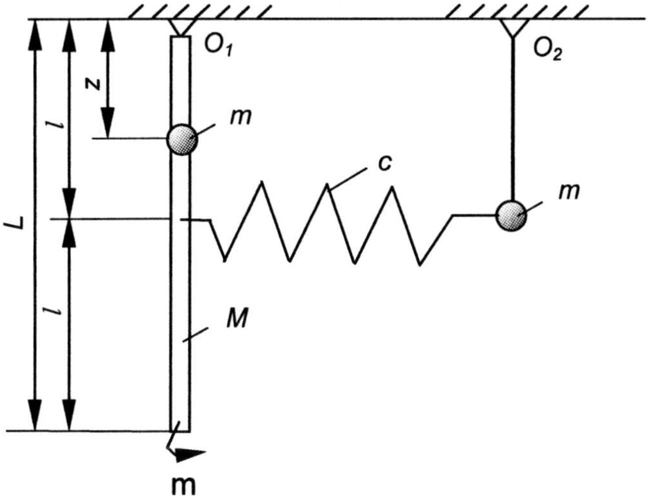


*Figure 9.10.*

9.6. A homogeneous disc has mass *M* and radius *R*. It rolls on the horizontal ground without sliding and it is linked to the pendulum with the spring (Figure 7.1.). The spring has rigidity *c*. Pendulum is hung to point $O_1$ and its mass is neglected. There are two concentrated masses (*m* and 2*m*) on the pendulum. Excitative force $F_\Omega = H \cdot \cos(\Omega t - \alpha)$ acts upon mass 2*m* (see Figure 9.11.). Estimate the system of differential equations. It is given: $l$=2[m]; *R*=1[m]; *c*=1000[N/m]; *M*=30[kg]; *m*=10[kg], the initial conditions: $t_0$=0[s]; $(\varphi_1)_0 = 0[rad]$; $(\varphi_2)_0 = 0[rad]$; $(\dot{\varphi}_1)_0 = 1[rad/s]$; $(\dot{\varphi}_2)_0 = 0{,}8[rad/s]$.

Final solution:

$$\frac{3}{2}MR^2\ddot{\varphi}_1 + 4cR^2\varphi_1 - 2Rcl\varphi_2 = 0$$

$$9ml^2\ddot{\varphi}_2 + 5mgl\varphi_2 + 2clR\varphi_1 - cl^2\varphi_2 = H \cdot 2l \cdot \cos(\Omega t - \alpha)$$

(Hint: See problem 7.1.)

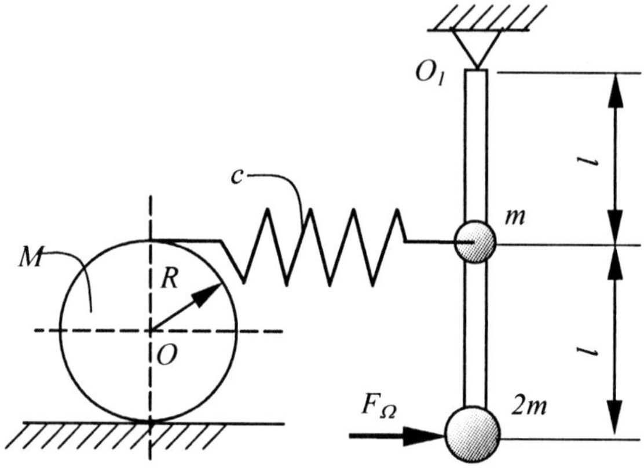


Figure 9.11.

# CHAPTER TEN

## TORSIONAL VIBRATIONS

### 10.1. Short Theory Review

These problems have the same procedure of solving as the problems described in the previous chapter. The torsional rigidity of a shaft ($c$) is used in the problems: $c=\frac{G\cdot I_0}{l}$, where $G$ is a shearing modulus of elasticity, $I_0$ is inertia's moment of shaft cross section, $l$ is the length of the shaft.

$I_0=\frac{d^4\pi}{32}$ - for the circular cross section, where $d$ is the diameter of the shaft, $I_0=\left(D^4-d^4\right)\frac{\pi}{32}$ - for the annular cross section, where $D$ is external and $d$ is the internal diameter of the shaft.

### 10.2. Problems

10.1. A homogeneous disc has mass $m$ and radius $R$. It is fixed for elastic shaft $AB$ in point $B$. Elastic shaft is fixed at the end (Figure 10.1.). It has the circular cross section with diameter $d$, shearing modulus of elasticity $G$, and inertia's moment of cross section $I_0$. Mass of the shaft is neglected. Estimate:

a) the differential equation of vibrations,
b) the circular frequency $\omega$,
c) the period $T$,
d) the law of torsional vibrations, if the initial conditions are: $t_0$=0[s]; $\theta_0$=0,087[rad]; $\dot{\theta}_0$=0[rad/s].

Solution:

a) It is possible to solve the problem using LaGrange's equation (2.9.). Kinetic energy of the system is (Figure 10.2.a)): $E_k=\frac{1}{2}J\cdot\dot{\theta}^2$,

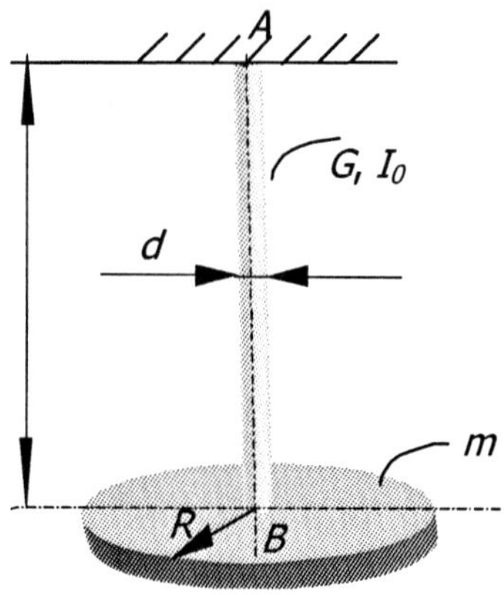


Figure 10.1.

where $J$ is inertia's moment of the disc for axis $AB$.
The potential energy (Figure 10.2.*a)*):

$$E_p = \frac{1}{2} c \cdot \theta^2,$$

where $c$ is torsional rigidity, and $c = \frac{G \cdot I_0}{l}$, $G$ is shearing modulus of elasticity, $I_0$ is inertia's moment of shaft cross section, $l$ is length of the shaft.

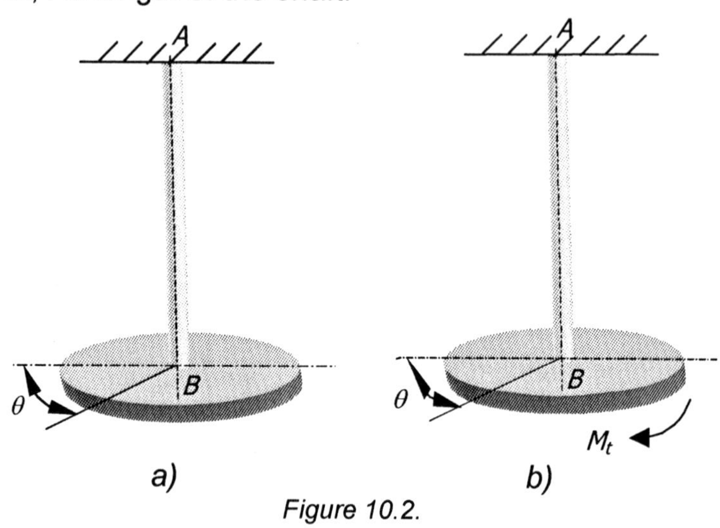


Figure 10.2.

LaGrange's equation (2.9.):

$$\frac{d}{dt}\left(\frac{\partial E_k}{\partial q}\right)-\frac{\partial E_k}{\partial q}+\frac{\partial E_p}{\partial q}=0$$

Put $E_k$ and $E_p$ in (2.9.):

$$J\cdot\ddot{\theta}-0+c\theta=0$$

$$J\cdot\ddot{\theta}+\frac{G\cdot I_0}{l}\theta=0$$

$$\ddot{\theta}+\frac{G\cdot I_0}{J\cdot l}\theta=0 \qquad \text{(10.2.1.)}$$

b) The circular frequency $\omega$ is:

$$\omega^2=\frac{G\cdot I_0}{J\cdot l}$$

$J=\frac{1}{2}mR^2$ - inertia's moment of the disc for axis $AB$

$I_0=\frac{d^4\pi}{32}$ - inertia's moment of shaft's cross section

$$\omega^2=\frac{2G\cdot d^4\pi}{32mR^2\cdot l}=\frac{G\cdot d^4\pi}{16mR^2\cdot l}$$

$$\omega=\frac{d}{4R}\sqrt{\frac{G\cdot\pi}{m\cdot l}}$$

c) The period: $T=\frac{2\pi}{\omega}=\frac{8\pi R^2}{d^2\sqrt{\frac{G\pi}{ml}}}$

d) It is possible to get the law of torsional vibrations using solution's form (2.3.):

$\theta=C_1\cos\omega t+C_2\sin\omega t$

$$\frac{d\theta}{dt}=\dot{\theta}=-\omega C_1\sin\omega t+\omega C_2\cos\omega t$$

Put the initial conditions in the expressions of $\theta$ and $\dot{\theta}$:

$0{,}087=C_1\cdot 1+C_2\cdot 0 \quad\Rightarrow\quad C_1=0{,}087[\text{rad}]$

$0=0+\omega\cdot C_2\cdot 1 \quad \Rightarrow \quad C_2=0[rad]$

The law is:

$$\theta = 0{,}087\cdot\cos\left(\frac{d^2}{4R}\sqrt{\frac{G\pi}{ml}}\right)$$

10.2. Solve the previous problem using the law of change of angular momentum.

Solution:

The law of change of angular momentum:

$J\ddot{\theta} = \sum M_{AB}$ ,

where $\sum M_{AB}$ is sum of the moments for axis *AB*. It is needed to swing the disc for angle $\theta$ (Figure 10.2.*b)*). Every torque in the same sense is positive. The system aspires to go back in the position of equilibrium (Figure 10.1.), so, the torsional moment $M_t$ is negative (Figure 10.2. *b)*).

$J\ddot{\theta} = -M_t$

The torsional moment $M_t$:

$M_t = c\cdot\theta$,

where $c = \dfrac{I_0\cdot G}{l}$ is the torsional rigidity.

$$J\cdot\ddot{\theta} = -\frac{G\cdot I_0}{l}\theta$$

$$J\cdot\ddot{\theta} + \frac{G\cdot I_0}{l}\theta = 0$$

$$\ddot{\theta} + \frac{G\cdot I_0}{J\cdot l}\theta = 0$$

The last differential equation is (10.2.1.) in the previous problem.

10.3. There are two discs fixed for elastic shaft *PQ* (Figure 10.3.). The left disc has radius $R_1$ and mass $m_1$, the right disc - radius $R_2$ and mass $m_2$. The diameter of the shaft is *d*. The mass of the shaft is neglected. Estimate the frequent equation and circular frequency of system's torsional vibrations. It is known: $m_1$=10[kg]; $m_2$=20[kg]; $R_1$=1[m]; $R_2$=1[m]; $l$=2[m]; d=0,05[m]; shearing modulus of elasticity $G=8{,}3\cdot 10^{10}$[N/m$^2$].

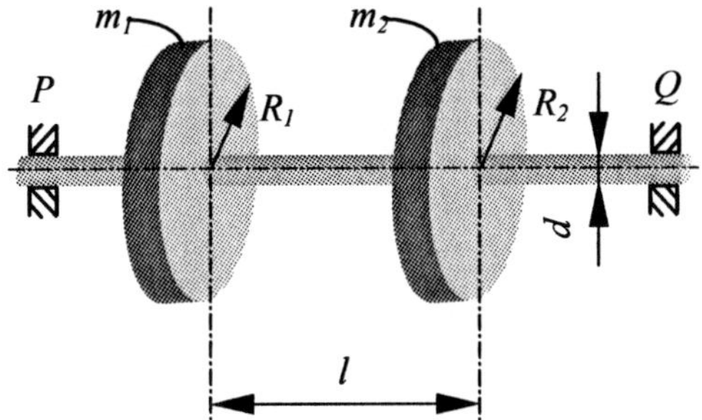


Figure 10.3.

Solution:
Kinetic energy of the system:

$$E_k = \frac{1}{2}J_1\dot{\varphi}_1^2 + \frac{1}{2}J_2\dot{\varphi}_2^2,$$

where $J_1$ is inertia's moment of the left disc (Table 7. of Appendix):

$$J_1 = \frac{1}{2}m_1R_1^2 = \frac{1}{2}10\cdot 1^2 = 5\,[\text{kgm}^2]$$

$J_1$ is inertia's moment of the right disc:

$$J_2 = \frac{1}{2}m_2R_2^2 = \frac{1}{2}20\cdot 1^2 = 10\,[\text{kgm}^2]$$

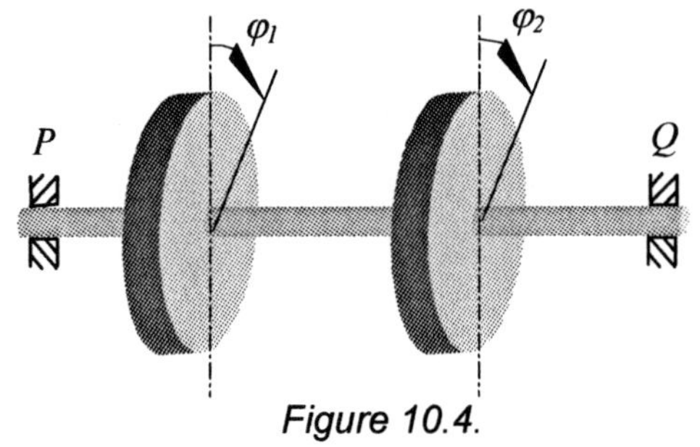


Figure 10.4.

Potential energy of the system: $E_p = \frac{1}{2}c(\varphi_2 - \varphi_1)^2$, where *c* is the torsional rigidity: $c = \frac{I_0 \cdot G}{l}$

$I_0$ is inertia's moment of cross section.

$I_0 = \frac{d^4 \pi}{32}$ for circular cross section

$I_0 = \frac{0{,}05^4 \pi}{32} = 0{,}614 \cdot 10^{-6}$ [m$^4$]

$$c = \frac{I_0 \cdot G}{l} = \frac{8{,}3 \cdot 10^3 \left[\frac{N}{m^2}\right] \cdot 0{,}614 \cdot 10^{-6} \left[m^4\right]}{2[m]} = 2{,}5481 \cdot 10^4 \left[Nm\right]$$

Put the expressions for $E_k$ and $E_p$ in LaGrange's equations (7.1.):

$$J_1 \ddot{\varphi}_1 + c(\varphi_2 - \varphi_1) \cdot (-1) = 0$$

$$J_2 \ddot{\varphi}_2 + c(\varphi_2 - \varphi_1) \cdot 1 = 0$$

$$J_1 \ddot{\varphi}_1 - c\varphi_2 + c\varphi_1 = 0$$

$$J_2 \ddot{\varphi}_2 + c\varphi_2 - c\varphi_1 = 0 \qquad \text{. . . . . . . . . . . . . (10.2.2.)}$$

The partial solutions have form (7.3.):

$\varphi_1 = A_1 \sin(\omega t + \beta)$

$\varphi_2 = A_2 \sin(\omega t + \beta)$

$\ddot{\varphi}_1 = -\omega^2 A_1 \sin(\omega t + \beta)$

$\ddot{\varphi}_2 = -\omega^2 A_2 \sin(\omega t + \beta)$

Put it in the system of differential equation (10.2.2.) and abbreviate the equations with sin($\omega t+\beta$):

$-\omega^2 A_1 J_1 - A_2 c + A_1 c = 0$

$-\omega^2 A_2 J_2 + A_2 c - A_1 c = 0$

$(c - \omega^2 J_1) A_1 - A_2 c = 0$

$-A_1 c + (c - \omega^2 J_2) A_2 = 0$

$$\Delta(\omega) = \begin{vmatrix} c - \omega^2 J_1 & -c \\ -c & c - \omega^2 J_2 \end{vmatrix} = 0$$

The frequent equation:

$$(c-\omega^2 J_1)(c-\omega^2 J_2)-c^2=0$$

$$J_1 J_2\omega^4-(cJ_2+cJ_1)\omega^2=0$$

Put the values for $J_1$, $J_2$ and $c$:

$$50\omega^4-3{,}82\cdot10^5\omega^2=0$$

The circular frequencies are:

$\omega_1$=0[s$^{-1}$] and $\omega_2$=87,43[s$^{-1}$].

10.4. The vibration's system is composed of two discs and two elastic shafts. Figure 10.5. shows the dimension of the system's parts. The masses of the shafts are neglected. Excitative moment $M=M_0\cdot\cos\Omega t$ acts upon the disc. Estimate the system of the differential equation, if it is given: $d_2=2\cdot d_1$; $m_1=2\cdot m_2$; $r_2=2\cdot r_1$, $G$.

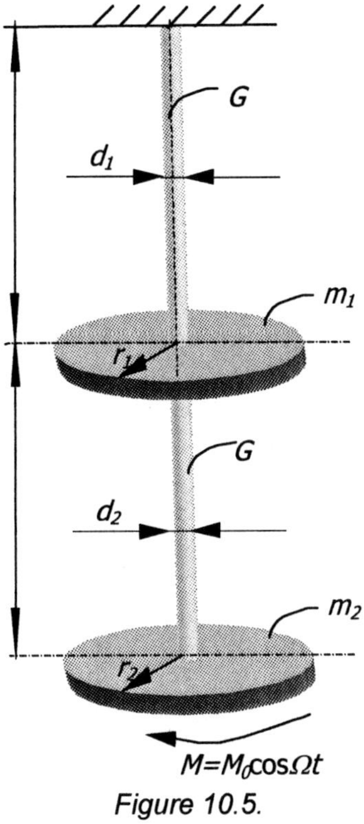


*Figure 10.5.*

Solution:

The kinetic energy:

$$E_k = \frac{1}{2} J_1 \dot{\varphi}_1^2 + \frac{1}{2} J_2 \dot{\varphi}_2^2$$

The potential energy:

$$E_p = \frac{1}{2} c_1 \varphi_1^{\ 2} + \frac{1}{2} c_2 (\varphi_2 - \varphi_1)^2$$

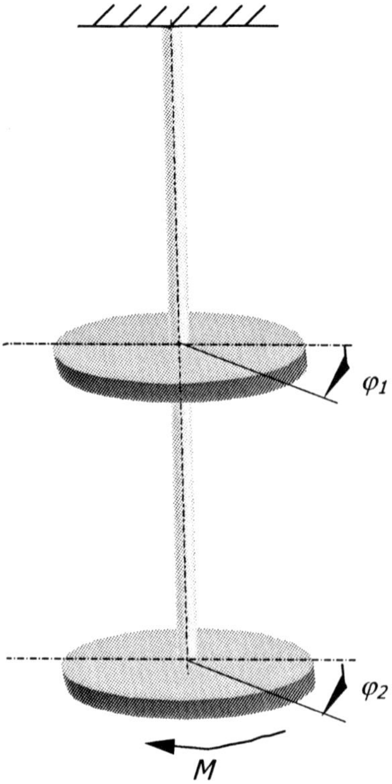


*Figure 10.6.*

LaGrange's equations:

$$\frac{d}{dt}\left(\frac{\partial E_k}{\partial \dot{\varphi}_i}\right) - \frac{\partial E_k}{\partial \varphi_i} + \frac{\partial E_p}{\partial \varphi_i} = Q_1 \qquad i=1,2$$

where Qi is generalized force (force or moment)
$Q_1$=0 and $Q_2$= $M=M_0 \cdot \cos \Omega t$

$$J_1 \ddot{\varphi}_1 - 0 + c_1 \varphi_1 + c_2(\varphi_2 - \varphi_1) \cdot (-1) = 0$$

$$J_2 \ddot{\varphi}_2 - 0 + c_2(\varphi_2 - \varphi_1) \cdot 1 = M_0 \cos \Omega t$$

$$J_1\ddot{\varphi}_1 + c_1\varphi_1 + c_2\varphi_1 - c_2\varphi_2 = 0$$
$$J_2\ddot{\varphi}_2 - c_2\varphi_1 + c_2\varphi_2 = M_0 \cos\Omega t$$

where $c_1 = \dfrac{G \cdot I_{01}}{l} = \dfrac{G \cdot D_1^4 \cdot \pi}{32 \cdot l}$; $\qquad J_1 = \dfrac{1}{2} m_1 r_1^2$

$$c_1 = \frac{G \cdot I_{02}}{l} = \frac{G \cdot d_2^4 \cdot \pi}{32 \cdot l} = \frac{G \cdot 16 d_1^4 \cdot \pi}{32 \cdot l} = \frac{G \cdot d_1^4 \cdot \pi}{2 \cdot l};$$

$$J_2 = \frac{1}{2} m_2 r_2^2 = \frac{1}{2}\frac{m_1}{2} 4 r_1^2 = m_1 r_1^2 .$$

10.5. Calculate the amplitudes of forced vibrations $A_1$ and $A_2$ for the vibration's system in the problem 10.4.

Final solution: $A_1 = \dfrac{\Delta_1}{\Delta} = \dfrac{c_2 \cdot M_0}{\left(c_1 + c_2 - J_1\Omega^2\right)\left(c_2 - J_2\Omega^2\right) - c_2^2}$

$$A_2 = \frac{\Delta_2}{\Delta} = \frac{\left(c_1 + c_2 - J_1\Omega^2\right) \cdot M_0}{\left(c_1 + c_2 - J_1\Omega^2\right)\left(c_2 - J_2\Omega^2\right) - c_2^2}$$

10.6. Write the differential equation of torsional vibrations and calculate the circular frequency $\omega$ for the vibration's system in Figure 10.7. Mass of the elastic shaft is neglected and it has a circular cross section with diameters $d_1$ and $d_2$.

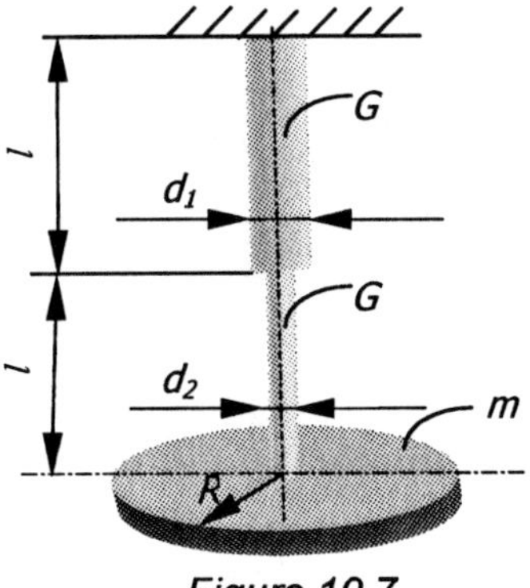


*Figure 10.7.*

Hint: It is possible to consider the elastic shaft like two torsional springs connected in series, and equivalent rigidity is: $\frac{1}{c_{ekv}} = \frac{1}{c_1} + \frac{1}{c_2}$ ($c_1 = \frac{I_{01} \cdot G}{l}; c_2 = \frac{I_{02} \cdot G}{l}$).

Final solution:

$$\ddot{\varphi} + \frac{c_{ekv}}{\frac{1}{2} mR^2} \cdot \varphi = 0$$

$$\omega = \sqrt{\frac{G\pi}{16 \cdot l \cdot m \cdot R^2 \cdot \left( \frac{1}{d_1^4} + \frac{1}{d_2^4} \right)}}$$

# APENDIX

| Table 1 | Beam supported at both ends |
|---|---|
| 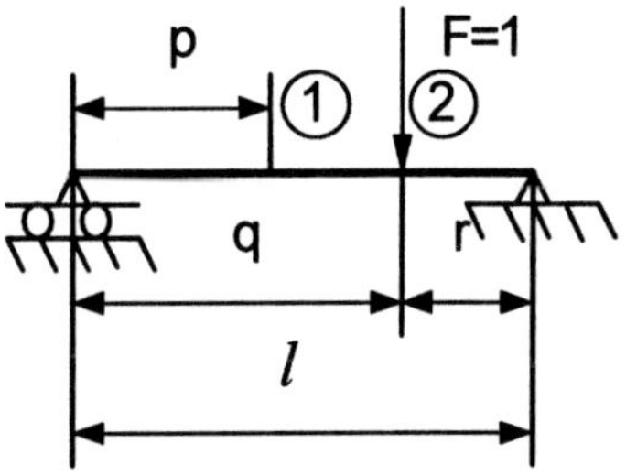 | $\alpha_{12} = \frac{l^3}{6EI}\left\{\frac{r}{l}\frac{p}{l}\left[1-\left(\frac{r}{l}\right)^2 - \left(\frac{p}{l}\right)^2\right]\right\}$ |
| 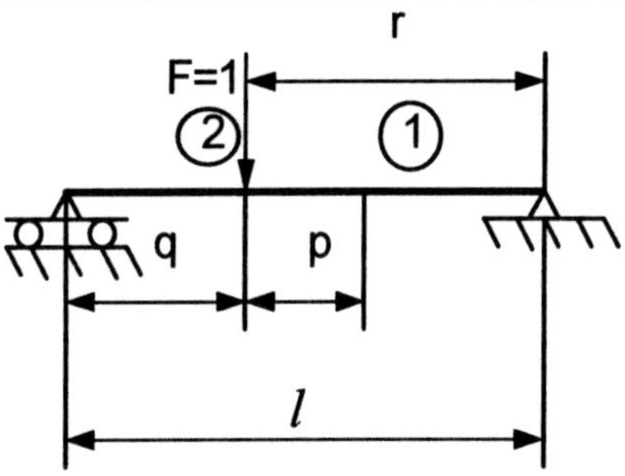 | $\alpha_{12}=\frac{l^3}{6EI}\left\{\frac{r}{l}\frac{q+p}{l}\cdot\left[1-\left(\frac{r}{l}\right)^2-\left(\frac{q+p}{l}\right)^2\right]\right\}$ |
| 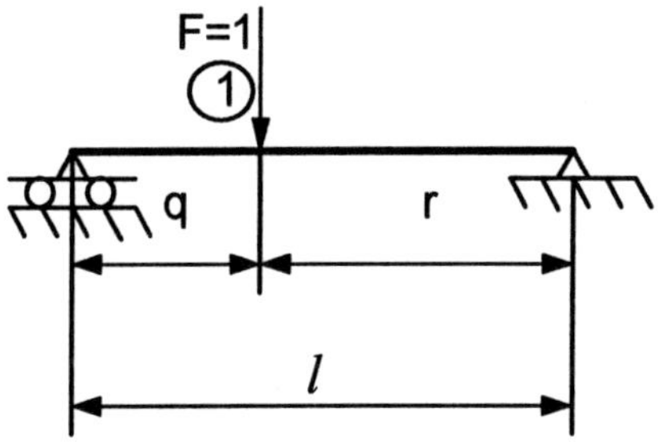 | $\alpha_{11} = \frac{l^3}{3EI}\left(\frac{q}{l}\right)^2\left(\frac{r}{l}\right)^2$ |

| Table 2 | Cantilever beam |
|---|---|
| 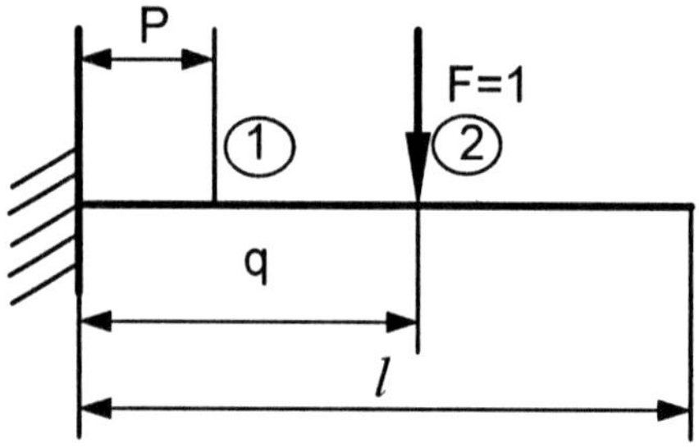 | $\alpha_{12} = \frac{l^3}{6EI}\left[3\frac{q}{l}\left(\frac{p}{l}\right)^2 - \left(\frac{p}{l}\right)^3\right]$ |
| 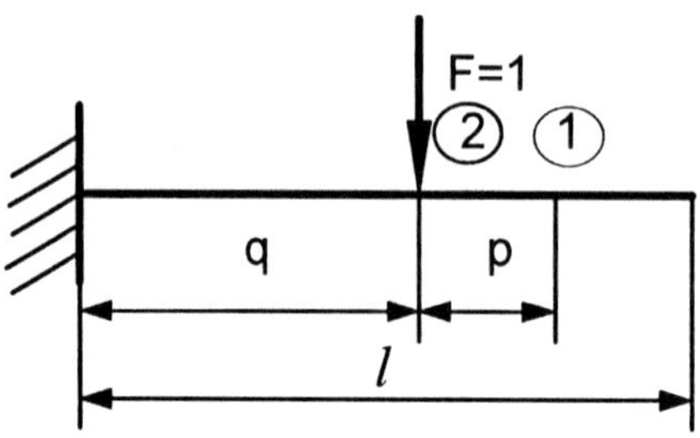 | $\alpha_{12} = \frac{l^3}{6EI}\left\{\left[3\frac{q}{l}\left(\frac{p+q}{l}\right)^2 - \left(\frac{p+q}{l}\right)^3\right] + \left(\frac{p}{l}\right)^3\right\}$ |
| 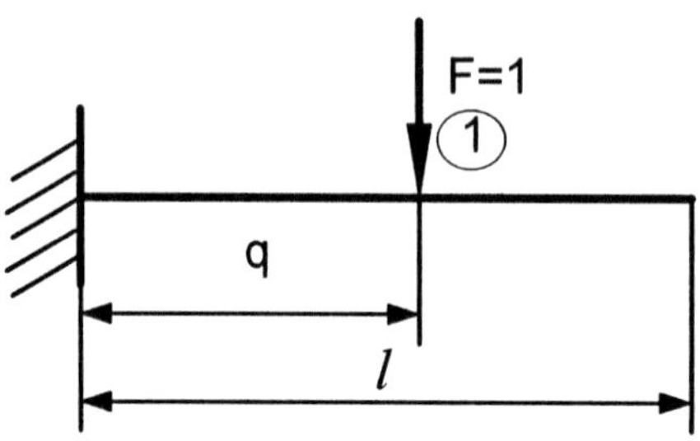 | $\alpha_{11} = \frac{q^3}{3EI}$ |

| Table 3 | Beam with an overhang |
|---|---|
| 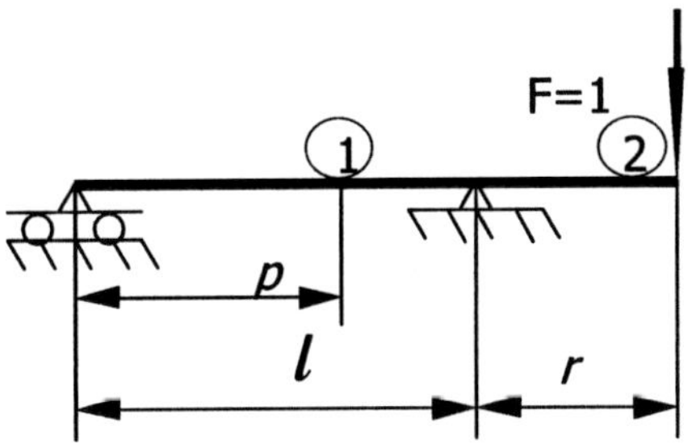  | $\alpha_{12} = -\frac{l^3}{6EI}\frac{r}{l}\left[\frac{p}{l} - \left(\frac{p}{l}\right)^3\right]$ |
| 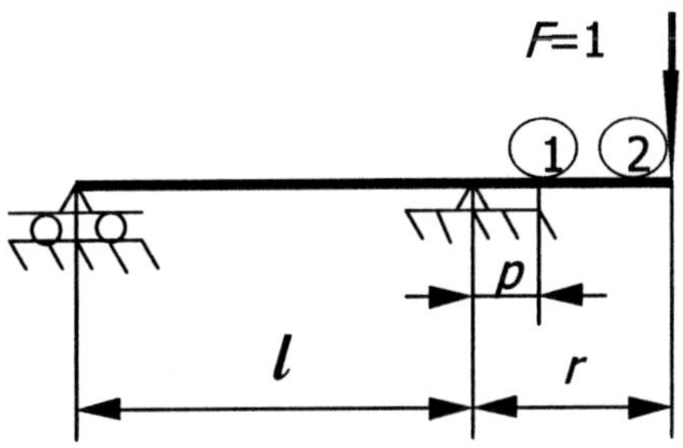  | $\alpha_{12} = -\frac{l^2 r}{6EI}\left[\left(1+\frac{p}{l}\right)^3 - 1 - \frac{p}{l} - \left(1+\frac{l}{r}\right)\left(\frac{p}{l}\right)^3\right]$ |
| 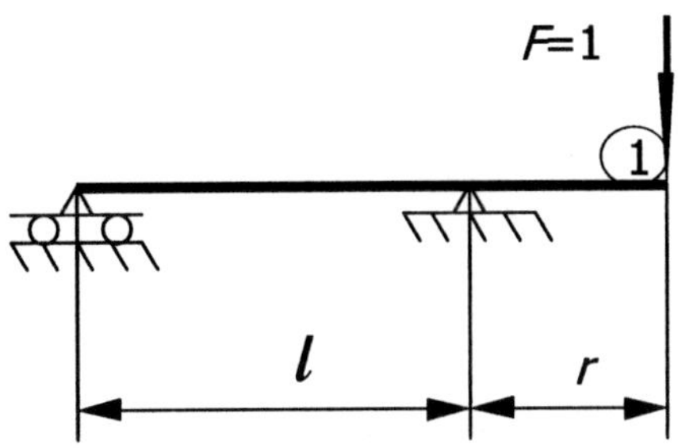  | $\alpha_{11} = \frac{r^2}{3EI}(l+r)$ |

| Table 4 | Statically indeterminate beams (beam with a fixation and the roller support) |
|---|---|
| 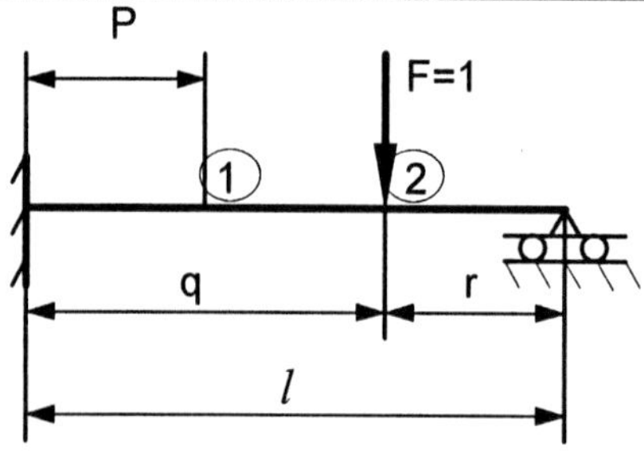 | $\alpha_{12}^{*}=\frac{l^3}{12EI}\frac{r}{l}\left(\frac{p}{l}\right)^2\left\{\left[3-\left(\frac{r}{l}\right)^2\right]\cdot\left(1-\frac{p}{l}\right)-2\left(\frac{r}{l}\right)^2\right\}$ |
| 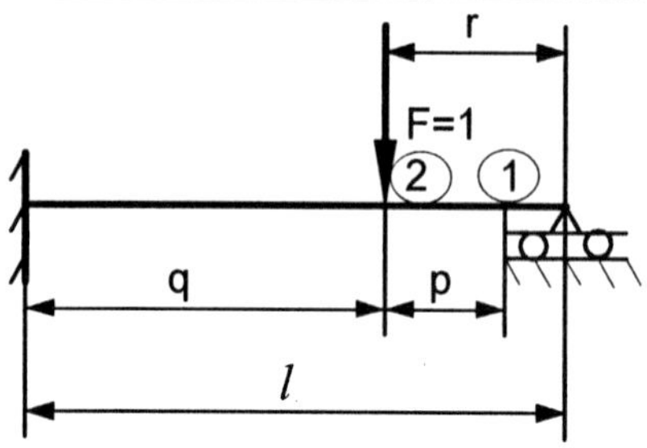 | $\alpha_{12}^{*}=\frac{l^3}{12EI}\left(\frac{q}{l}\right)^2\left(1-\frac{q+p}{l}\right)\cdot\left[3\frac{r}{l}-\left(2+\frac{r}{l}\right)\cdot\left(1-\frac{q+p}{l}\right)^2\right]$ |
| 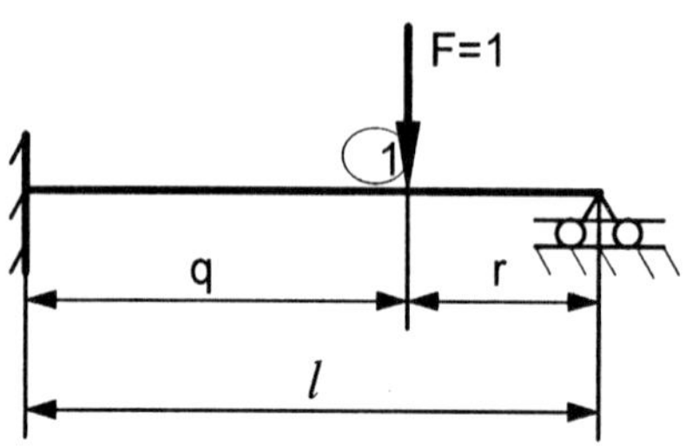 | $\alpha_{11}^{*}=\frac{l^3}{12EI}\left(\frac{q}{l}\right)^3\left(\frac{r}{l}\right)^2\left(3+\frac{r}{l}\right)$ |

| Table 5 | Statically indeterminate beams (beam with a fixation at both ends) |
|---|---|
| 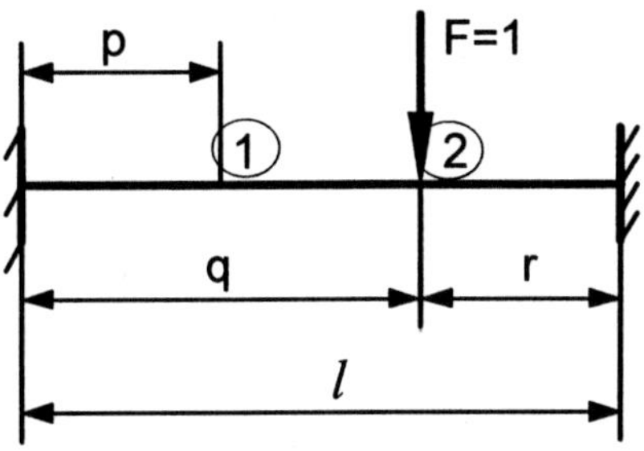  | $\alpha_{12}^{*}=\frac{l^3}{6EI}\left(\frac{r}{l}\right)^2\left(\frac{p}{l}\right)^2\left[3\frac{q}{l}-\right.$ $\left.-\left(\frac{r}{l}+3\frac{q}{l}\right)\frac{p}{l}\right]$ |
| 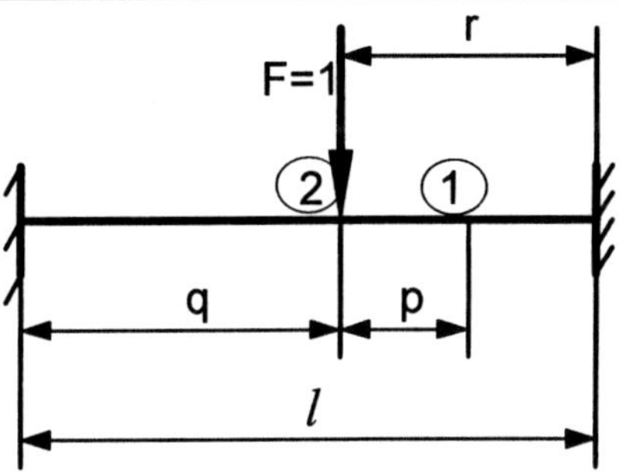  | $\alpha_{12}^{*}=\frac{l^3}{6EI}\left(\frac{q}{l}\right)^2\left(1-\frac{q+p}{l}\right)^2\cdot$ $\cdot\left[3\frac{r}{l}-\left(\frac{q}{l}+3\frac{r}{l}\right)\frac{r-p}{l}\right]$ |
| 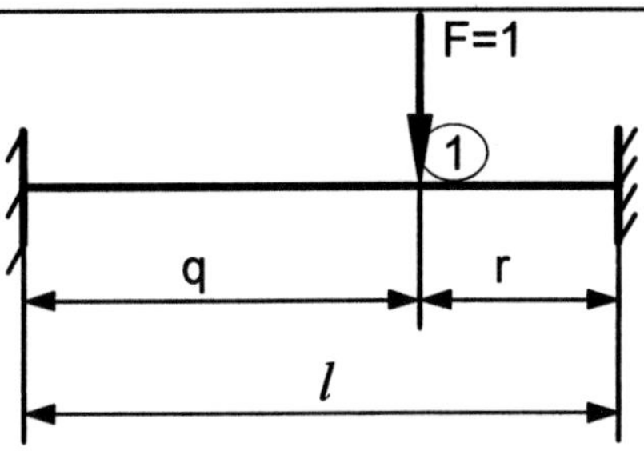  | $\alpha_{11}^{*}=\frac{l^3}{3EI}\left(\frac{q}{l}\right)^3\left(\frac{r}{l}\right)^3$ |

# Table 6. Review of some expressions for kinetic and potential energy

| Figure | Description | $E_k$ | $E_p$ |
|---|---|---|---|
| 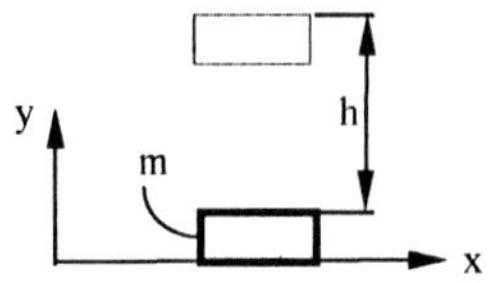 | Rectilinear vertical motion upwards of a concentrated mass | $E_k = \frac{1}{2} m\dot{y}^2$ | $E_p = mgh$ |
| 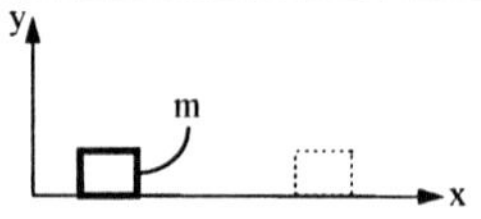 | Rectilinear horizontal motion of a concentrated mass | $E_k = \frac{1}{2} m\dot{x}^2$ | $E_p = 0$ |
| 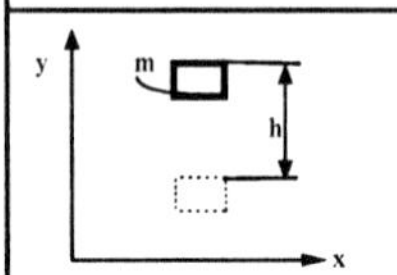 | Rectilinear vertical motion downwards of a concentrated mass | $E_k = \frac{1}{2} m\dot{y}^2$ | $E_p = -mgh$ |
| 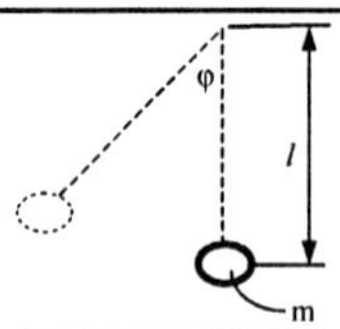 | Circular motion upwards of a concentrated mass | $E_k = \frac{1}{2} m(l\dot{\varphi})^2$ | $E_p = mg(l - l\cos\varphi)$ |
| 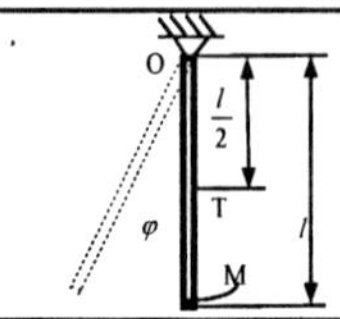 | A homogeneous lever with mass *M*, hooked up at the end | $E_k = \frac{1}{2}\left(\frac{1}{3} Ml^2\right)\dot{\varphi}^2$ | $E_p = Mg\left(\frac{l}{2} - \frac{l}{2}\cos\varphi\right)$ |
| 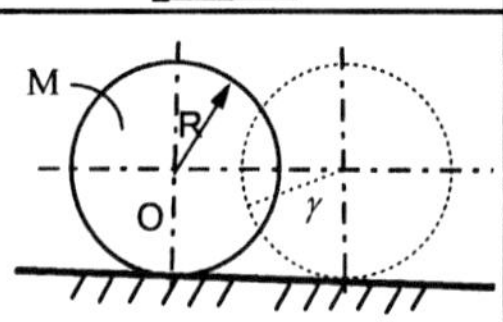 | Plane motion of a homogeneous disc (rolling) on the horizontal ground | $E_k = E_{ktr} + E_{krot} = \frac{1}{2} Mv_0^2 + \frac{1}{2}\left(\frac{1}{2} MR^2\right)\dot{\gamma}^2$ | $E_p = 0$ |

Table 7. Review of some expressions for the moment of inertia for axis

| Shape | Figure | Description | Moment of inertia with respect to an axis *x* |
|---|---|---|---|
| Ball (mass *m*, radius *r*) | 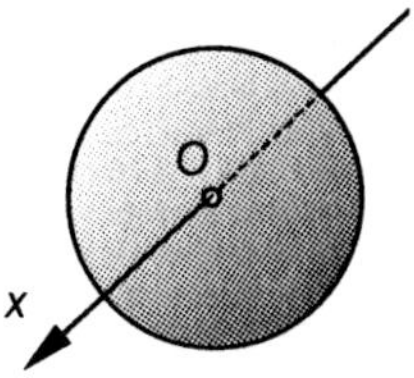 | The axis *x* is in the center of the ball | $J_x = \frac{2}{5} mr^2$ |
| Disk (mass *m*, radius *r*) | 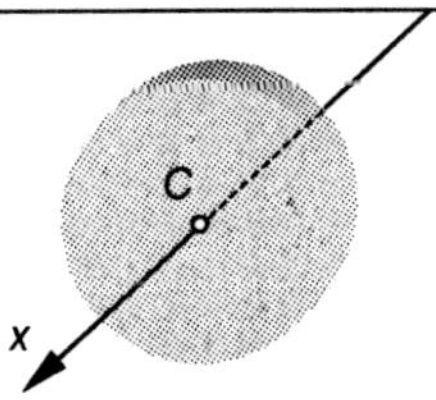 | The axis *x* is in the disk's center of gravity | $J_x = \frac{1}{2} mr^2$ |
| Disk (mass *m*, radius *r*) | 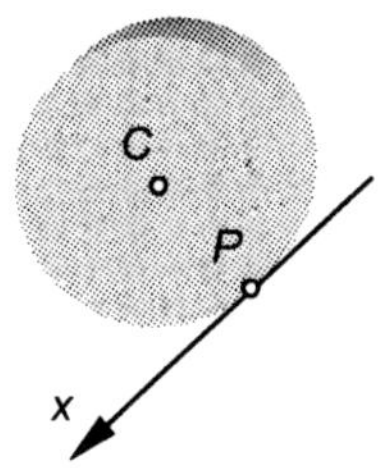 | The axis *x* is in the edge of disk | $J_x = \frac{3}{2} mr^2$ |
| Bar (mass *m*) | 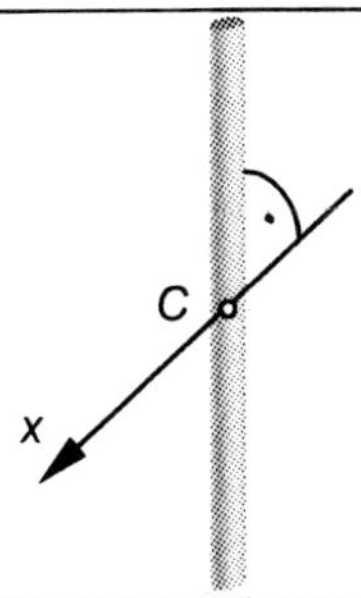 | The axis *x* is in the bar's center of gravity | $J_x = \frac{1}{12} ml^2$ |
| Bar (mass *m*) | 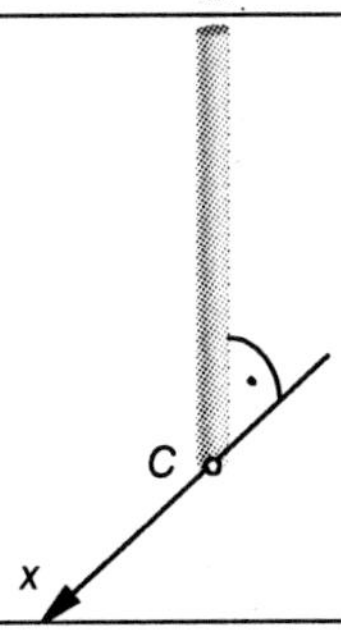 | The axis *x* is in the boundary point of bar | $J_x = \frac{1}{3} ml^2$ |

Table 8. Review of units interdependence

| Units of length | |
|---|---|
| 1"=1 in (inch) = 0,0254 m (metre) | 1 m =39,37 in |
| 1 ft (foot) = 0,3048 m | 1 m =3,28084 ft |
| 1 yd. (yard) =0,9144 m | 1 m =1,09361 yd. |
| **Units of mass** | |
| 1 lb. (pound) = 0,45359 kg (kilogramme) | 1 kg =2,20462 lb. |
| 1 ton (2240 lbs) = 1016,05 kg | 1t(1000kg)=0,9842ton |
| **Units of stress** | |
| 1 ton per sq. in.=15444296,952Pa $\left(\frac{N}{m^2}\right)$ | 1Pa=$6,47\cdot 10^{-8}$ ton per sq.in. |
| 1 lb. per sq. in. = 6894,761 Pa | 1 Pa = $1,45\cdot 10^{-4}$ lb. per sq.in. |
| 1 lb. per sq. ft. = 47,88097 Pa | 1 Pa = 0,02088 lb. per sq.ft. |

# Literature

[1] Ferdinand P. Beer, E. Russell Johnston, Jr.: "Mechanics for Engineers: Statics and Dynamics", 2000.

[2] Draca,K., Salai,I., Cigic,A. - “Oscilacije” - Faculty of Mechanical Engineering, University in Mostar, Mostar, 1991.

[3] Baranenkov,G.S., Demidovic,B.P. i drugi autori - “Zadaci i rijeseni primjeri iz vise matematike” - Tehnicka knjiga, Zagreb,1978.

[4] Targ,S.M. - “Kratkii kurs teoreticeskoi mehaniki”, Moskov, 1961.

[5] Cigic,A. - “Mehanika, zbirka rijesenih ispitnih zadataka” - Faculty of Mechanical Engineering, University in Mostar, Mostar, 1990.

[6] Mescerski,I.V. - “Sbornik zadac po teoreticeskoi mehaniki”, Moskov, 1956.god.

[7] Pasic,H. - “Statika”, Svjetlost, Sarajevo, 1987.

[8] Raskovic,D. - “Tablice iz otpornosti materijala”, Gradjevinska knjiga, Beograd,1987.

[9] Dolecek,V. - “Zbirka zadataka iz dinamike i oscilacija”, Svjetlost, Sarajevo, 1981.

[10] Karabegovic, I. - “Tehnicka mehanika1 - statika”, University in Sarajevo, Sarajevo, 1996.

[11] Karabegovic, I. - “Tehnicka mehanika 2 - kinematika”, University in Sarajevo, Sarajevo, 1994.

Printed in the United States
34601LVS00003B/245

9 781931 768429